Get Sales Faster

CHAD JOHNSON

A guide to help close deals faster for people who don't sell online.

Get Sales Faster

Requests for permission should be addressed to:
chad@personaldevelopmentworkshop.com

This book may be purchased for educational, business, or sales promotional purposes.

Get Sales Faster
All Rights Reserved.
Copyright © 2023 Chad Johnson
v1.0
ISBN: 978-1-304-80103-6

The opinions expressed in this manuscript are solely the opinions of the author and do not represent the opinions or thoughts of the publisher. The author has represented and warranted full ownership and/or legal right to publish all the materials in this book.

This book may not be reproduced, transmitted, or stored in whole or in part by any means, including graphic, electronic, or mechanical, without the express written consent of the publisher except in the case of brief quotations embodied in critical articles and reviews.

Cover Design © 2023 Chad Johnson. All rights reserved – used with permission.

PRINTED IN THE UNITED STATES OF AMERICA

Contents

Preface. i
Introduction. 1
The Answers are in the Fundamentals. 4
Know Thy Customer. 26
Get Better Meetings. 36
Find a Needle in a Needle Stack. 42
Be a Market Maker. 57
6 Deadly Sins of Sales. 77
6 Essential Skills to Increase Your Sales Success. 99
Conclusion: Connecting the Dots 120

Get Sales Faster

Preface.

Welcome to "Get Sales Faster," a book where traditional wisdom meets modern sales strategies, rooted in the principle that true success is not a quick sprint, but a marathon built on persistence, strategy, and understanding. This book comes on the heels of "Get Good at Sales" and is a prequel to "Get Better at Sales." I encourage you to read these books to get more insight into the ideas laid out here in a more condensed fashion.

At its core, this book focuses on the complex sale. It emphasizes the importance of building relationships and crafting business cases that resonate with your target customers. Tailoring solutions to specific problems encourage you to sell a product and offer a solution that adds genuine value.

The sales profession is fast-paced; the allure of quick fixes and rapid results often seduces us. However, this book is an invitation to embrace a counterintuitive philosophy: slowing down to accelerate success. It is a guide crafted from the belief that real progress, especially in complex sales environments, is a product of meticulous preparation, planning, and a deep understanding of your customer.

"Get Sales Faster" is built on the foundation of the 'slight edge philosophy.' It's a reminder that the path to becoming an overnight success is, paradoxically, a long and steady one. Each chapter of this book is designed to guide you through the nuances of a sophisticated sales process. From mastering the fundamentals to deeply understanding your customer, from securing better meetings to becoming a market maker, this book encompasses the essential components of the sales journey.

"Slight Edge" by Jeff Olson is a motivational book that focuses on the power of small, consistent actions over time to create significant life changes. Olson emphasizes that success is not the result of one-time significant actions but rather

the culmination of daily small decisions and steps. He argues that these seemingly insignificant actions, when compounded over time, can lead to massive success.

This concept is not new, but it worked for me even before I knew about or understood the slight edge philosophy. I understood that people need to understand before they buy anything. And sometimes it takes some people longer to understand what you are trying to do than others. Some may never get it. That's ok! Now, you must improve at explaining what you do, especially in the complex sales arena.

In this book, we will explore the concept of proactive and compounding actions. It's about making consistent, strategic moves that may initially seem small but lead to significant long-term gains. It emphasizes building relationships and crafting business cases that resonate with your target customers, tailoring solutions to specific problems. "Get Sales Faster" underscores that the journey to becoming an overnight success is paradoxically long and steady. Each chapter is meticulously crafted, guiding you through various aspects of the sales process, from understanding fundamentals to deeply knowing your customer, securing better meetings, and becoming a market maker. You will explore topics such as identifying and qualifying the right customers, understanding the '6 Sins of Sales,' and mastering the '6 Essential Sales Skills.' Each chapter blends theory and actionable insights, driving you toward sales excellence.

"Have you ever wondered why some sales strategies fail despite seeming effective on the surface?"

Often, we come across methods and techniques that are widely acclaimed, promising quick results and high conversion rates. Yet, the reality of their effectiveness can be quite different. The primary reason for this discrepancy lies in the complexity and uniqueness of each sales situation. A strategy that works wonders in one context may falter in another due to various factors such as customer needs, market dynamics, and the specific nuances of a product or service.

The success of a sales strategy is not just about the tactics employed; it's rooted in the relationship and trust built with the customer. A strategy that overlooks the importance of understanding the customer's unique challenges, goals, and decision-making processes is often doomed to fail, regardless of how successful it appears on paper. Additionally, many strategies focus on short-term gains rather than building long-term relationships, leading to a lack of sustainable success.

Therefore, while a sales strategy might appear effective at first glance, its true measure of success lies in its adaptability, customer-centric approach, and ability to foster lasting relationships. It's essential to look beyond the surface and critically evaluate how a strategy aligns with the specific needs and values of the salesperson and the customer to understand why some strategies succeed while others do not."

"Do you ever ponder what differentiates a truly successful salesperson from one who consistently struggles despite seemingly using similar techniques?"

We discuss this concept as the A player. The goal is to play like an A player. Several key factors come to light when considering what differentiates a successful salesperson from one who consistently struggles despite seemingly using similar techniques.

Firstly, the ability to genuinely connect with clients plays a crucial role. Successful salespeople often possess a deep empathy and understanding of their clients' needs, fears, and desires, allowing them to tailor their approach in a way that resonates on a personal level.

Next, they are relentless in pursuing a deal and do what it takes to get it done. These actions are consistently performed and go beyond what any of their competitors will do. These actions are noticed by their prospects and customers, and they are regularly chosen over their competition because of it.

Additionally, successful salespeople exhibit a high degree of adaptability and learning. They quickly learn from their experiences, adapting their strategies to suit different situations and clients better. This contrasts with struggling salespeople who might rigidly adhere to specific techniques without considering their effectiveness in varying contexts.

Lastly, successful salespeople often deeply understand their product or service and the market in which they operate. This knowledge goes beyond surface-level features and benefits; it encompasses understanding how their offering fits into the broader context of their clients' lives or businesses.

"Have you considered the possibility that sometimes the most promising deals fall through not because of the strategy itself but due to overlooked details or rushing the process?"

It's an intriguing thought that calls for a deeper examination of how we handle sales interactions. Haste makes waste! Often, in our eagerness to close deals quickly, we

might skip over crucial details or fail to build a sufficiently deep understanding of our client's needs and concerns. This oversight can lead to a misalignment between what we offer and what the client truly requires or values.

Rushing the process can also mean missing out on building a solid relationship foundation with the client. Sales, at its core, is about trust and rapport. When we move too fast, we risk not fully establishing this trust, which is essential for a client to feel confident and comfortable in making a decision. Furthermore, in complex sales environments, decision-making often involves multiple stakeholders. Rushing might lead to not engaging all these stakeholders effectively, leaving out key influencers who play a crucial role in the decision-making process.

While a proactive and efficient approach is valuable, it is equally important to be thorough and attentive to detail. Balancing speed with careful consideration ensures that promising deals are nurtured to their full potential, reducing the risk of them falling through due to avoidable oversights or haste.

The devil is in the details. At the same time, if your client is unwilling to make the time for the deal, they aren't interested, or you're speaking with the wrong person, Slowing the process down and working out the details will save you time in the long run because if they don't want to work to move the deal forward, you can stop wasting your time with them.

Be Prepared

To get sales faster, you must be prepared and work on the fundamentals daily. The concept of preparedness resonates deeply with my philosophy. It reminds me of a saying that resonates with many successful people: being an "overnight success 20 years later." This encapsulates the essence of what I advocate in "Get Sales Faster." Success in sales, as in many areas of life, is often the result of long-term, consistent effort and preparation. The journey to mastery and success is not achieved in a flurry of hasty decisions and actions but through years of learning, adapting, and refining our approach. This book offers insights and strategies honed over years of experience, which may seem like an overnight success to the outside world but are indeed the fruits of persistent, dedicated work.

I want to emphasize that this is a quick guide for a not-so-quick process to help get sales faster and understand the philosophy of being prepared and resonating with your clients based on your personality traits and qualities. Although counterintuitive, spending more time on a deal will get you the deal faster because the deal will be ironclad.

Here is what we will cover in this book.

Chapter Overview

1. **The Answers are in the Fundamentals**: Focuses on the foundational elements of sales. It emphasizes the importance of mastering basic skills and principles as the cornerstone of successful selling.
2. **Know Thy Customer**: Looks into the importance of deeply understanding your customers. This chapter provides strategies for identifying customer needs, preferences, and decision-making processes, enabling a more tailored and effective sales approach.
3. **Get Better Meetings**: Offers insights and tactics on securing and conducting productive and meaningful client meetings. It covers everything from setting agendas to ensuring client engagement and follow-up.
4. **Find a Needle in a Needle Stack**: Discusses advanced techniques for identifying high-value prospects in a crowded market. This chapter focuses on refining targeting strategies to improve efficiency and effectiveness in sales.
5. **Be a Market Maker**: Explores innovative ways to create and capitalize on new market opportunities. This chapter encourages thinking outside the box and leveraging unique selling propositions to stand out in the market.
6. **6 Deadly Sins of Sales**: Highlights common pitfalls and mistakes in sales and how to avoid them. This chapter serves as a cautionary guide to help readers navigate and overcome potential challenges in their sales journey.
7. **6 Essential Skills to Increase Your Sales Success**: Details the critical skills and competencies needed to excel in sales. From communication and negotiation to adaptability and resilience, this chapter is a roadmap to developing the traits of a top salesperson.

Reflection

Now, let me share a personal story that has shaped my approach to sales. Once, just once, in my eagerness to close a deal quickly, I overlooked the involvement of a critical stakeholder in my client's decision-making process. Despite my preparedness and confidence, the deal was unexpectedly shot down. This key person, overlooked in my haste, had significant influence and insufficient knowledge about the proposed solution. This experience was a pivotal lesson in the importance of not rushing the sales process. I lost a valuable opportunity by not allowing enough time for all stakeholders to consider the information and be involved.

This incident forms the essence of my approach and this book. It teaches the value of slowing down the buying process – adding an extra meeting, involving more decision-makers, and presenting comprehensive information. Though seemingly

time-consuming, such steps filter out clients who aren't ready for a decision based on thorough information. It's about understanding the delicate balance between being proactive and patient, quick yet thorough, assertive yet considerate.

"Get Sales Faster" isn't just about selling quickly; it's about selling right. It's about delayed gratification, understanding when to speed up and when to slow down, all while keeping your customer's best interests at heart. This approach will foster better relationships and ensure more successful and sustainable sales in the long run.

This book is not just about selling; it's about excelling in the art of meaningful and effective salesmanship. Let's embark on this journey together, where each step, no matter how slight, leads you closer to your goal: to get sales faster, smarter, and more efficiently than ever before.

Introduction.

Complex Sales is not a monolithic entity but a multifaceted discipline encompassing various roles, strategies, and approaches. Whether you're a direct salesperson building client relationships, a digital marketer crafting persuasive campaigns, or a startup founder pitching to investors, you are engaged in selling. Sales is a vast concept, and its implications are profound. It's not just about closing deals; it's about understanding human psychology, building trust, and solving problems. It's about adaptability, resilience, and the pursuit of excellence. It's about the art of communication, negotiation, and influence.

Complex sales processes could be faster, more convenient, and have long sales cycles. Everyone on the selling side wants sales faster. Everyone on the buying side wants to spend slower. How do you get sales faster? Getting sales faster is your problem to solve. You can't make anyone buy anything. People buy on their terms, and getting your prospects to turn to customers can, at times, be a difficult position. To understand complex sales, you need to know what they are. Complex sales processes are many things, but what they are not is a person perusing the internet and clicking purchase.

In professional sales, there are two kinds of deals. Transactional and Complex.

Transactional Sales

Transactional Sales: *Transactional sales, also known as low-touch sales, refer to selling relatively low-cost, standardized products or services with a primary emphasis on efficiency and speed. The goal in transactional sales is to complete the deal quickly, often in a single interaction, and move on to the next customer. This sales approach is typically associated with industries where customers have immediate needs and can make purchase decisions based on readily available information.*

Key Characteristics:

- **Low Cost**: Transactional sales typically involve relatively inexpensive products or services with standardized features.
- **Short Sales Cycle:** The sales cycle is brief, often measured in minutes to hours, as the customer's decision-making process is straightforward.
- **Limited Customer Engagement**: Interactions with customers are usually short, focusing on closing the deal based on price and features.
- **Small Decision-Making Unit (DMU)**: The DMU typically consists of one or a few individuals, such as procurement professionals or end-users.
- **Product-Centric Approach**: Salespeople emphasize the product's features and benefits and may use pricing and discounts to close deals.
- **Low Customer Loyalty**: Customer loyalty may be lower because customers are likelier to switch providers based on price or convenience.
- **Metrics**: Metrics in transactional sales often revolve around sales volume, conversion rates, and average transaction value.

Examples: Online retail, fast-food restaurants, basic software subscriptions, and consumer electronics stores often use transactional sales approaches.

Complex Sales

Description: Complex sales, or high-touch sales, involve selling high-value, customizable, or specialized products or services that require a consultative and relationship-focused approach. These sales processes are characterized by longer and more involved customer interactions, often spanning weeks to months or even years. Complex sales are expected in industries where the customer's needs are diverse and unique, and solutions must be tailored to meet specific requirements.

Key Characteristics:

- **High-Value Offerings**: Complex sales deal with products or services that are high in value and often require customization to meet the customer's specific needs.
- **Extended Sales Cycle**: The sales cycle involves multiple stages, including needs assessment, solution development, and negotiation.
- **Intensive Customer Engagement**: Sales professionals engage in deep, ongoing customer interactions, building relationships and understanding their complex needs.
- **Large Decision-Making Unit (DMU)**: The DMU includes various stakeholders, such as decision-makers, influencers, technical experts, and executives, each with their priorities and concerns.

- ❖ **Solution-Oriented Approach**: Salespeople focus on understanding the customer's challenges and developing tailored solutions that address those challenges.
- ❖ **High Customer Loyalty**: Building strong relationships and delivering customized solutions can increase customer loyalty in complex sales.
- ❖ **Metrics**: Metrics for complex sales may include customer satisfaction, customer lifetime value, and the sales cycle length.
- ❖ **Examples:** Enterprise software solutions, industrial equipment, consulting services, and healthcare solutions often require complex sales processes due to their high value and customization requirements.

Notice the glaring difference? Transactional sales require many deals on small dollar items. Complex sales require fewer sales with higher dollar items. Another issue is that salespeople try to make the complex easy or approach complex sales as a transactional process. No bueno.

Getting sales faster in a complex sales process with long sales cycles can be challenging. Using these strategies, you can help speed up the sales process and close deals more quickly, even in complex sales environments with long sales cycles.

In reflecting upon my sales experiences, one key realization stands out: the duration of a deal, in terms of calendar time, might remain the same whether we rush or take it slow. However, the actual time and effort invested in the deal can vary significantly. This distinction is crucial. A deal nurtured over time, with careful attention to every detail and stakeholder, might span the same number of weeks as a rushed deal. Yet, the quality of engagement, the depth of understanding, and the strength of the relationships built vastly differ. This thoughtful approach ensures that every moment spent on the deal adds more value, creating a robust and well-informed sales process.

The concept of getting sales faster in complex deals is counterintuitive. Sales in a complex sales process take time, which can and will take as long as it takes. The ideas in this book will help you navigate the complex sale, redefine your process, and focus your efforts. These principles will get you sales faster because you aren't wasting time on opportunities that aren't opportunities. Sometimes, you need to slow the process down to get the deal. Some steps in your sales process you will find cannot be skipped. Skipped steps to get sales faster will ultimately lead to the last sales. Getting sales faster doesn't necessarily mean shortening the timeframe of the sales process, and adding an extra meeting doesn't necessarily mean extending the time frame it takes to close the deal.

Chapter 1

The Answers are in the Fundamentals.

We will always work on the fundamentals. You can't get better, let alone good at sales, if you aren't working on the fundamentals.

There are so many books focused on closing. Sales training is focused on product knowledge and knowing how to position your customers to buy based on your product features. After a rep has been with a company for so long, they know their products so well and what they can do that they can sell them to anyone. The problem is they don't know how to get in front of the customers to tell them to.

Too much time is spent on product/company awareness, and marketing sucks. More time is spent on positioning yourself to get in front of the right people to speak with initially. View this section as a guide to help you close business sales faster for people who don't sell online transactional sales.

You do need to know how to close, you do need product knowledge, and you do need to know about yourself. These aspects of the sales process are essential, but selling goes way beyond that, especially in the complex sales process.

The fundamentals of sales are the foundational principles and skills that sales professionals rely on to excel in their roles. These fundamentals serve as the building blocks for success in sales.

Sales Fundamentals

1. **Understanding the Customer:** Sales professionals must thoroughly understand their target audience, including their needs, preferences, pain points, and buying behavior. This knowledge helps in tailoring solutions to meet customer needs effectively.
2. **Effective Communication:** Sales professionals need strong communication skills to convey the value of their products or services clearly and persuasively. This includes active listening, asking probing questions, and articulating benefits concisely.
3. **Building Relationships:** Building and maintaining solid customer relationships is crucial for long-term success. Trust and rapport are the cornerstones of lasting customer relationships.
4. **Prospecting:** Identifying and targeting potential customers or leads is fundamental to sales. Effective prospecting involves finding individuals or organizations likely to be interested in the product or service.
5. **Time Management:** Salespeople often have a range of tasks to juggle, from lead generation to follow-ups. Time management skills are essential to ensure valuable time is spent on the most productive activities.
6. **Sales Process Knowledge:** Understanding the stages of your sales process and how to navigate them is crucial.
7. **Adaptability:** The sales landscape can change rapidly. Being adaptable and open to new approaches, technologies, and market shifts is critical to staying competitive.
8. **Resilience:** Rejection and setbacks are common in sales. Resilience and a positive mindset are essential for bouncing back and maintaining motivation.
9. **Continuous Learning:** Successful sales professionals always continue learning. They stay updated on industry trends, sales techniques, and product developments to remain at the top of their game.
10. **Ethical Conduct:** Ethical behavior is essential in sales. Trust is easily damaged, so maintaining honesty and integrity in all interactions is crucial for long-term success.
11. **Goal Setting:** Sales professionals often work with targets and quotas. Setting clear, achievable goals helps in maintaining focus and measuring success.
12. **Follow-up and Customer Service:** Providing excellent post-sale service and following up with customers ensures satisfaction and can lead to repeat business and referrals.

These fundamentals are the foundation for sales success, whether in traditional face-to-face sales, inside sales, B2B, B2C, or any other sales environment. Sales professionals who master these fundamentals are well-positioned to excel in their roles and build rewarding careers in sales

Sales Strategy

Strategy is a way to stay two steps ahead of the sale. The goal of a strategy is to execute the objectives by establishing the following steps to further your deal.

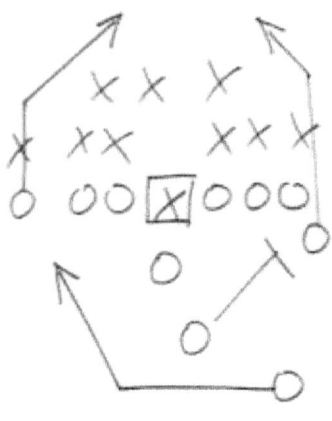

Developing an effective sales strategy is crucial for achieving consistent and sustainable revenue growth. It involves a combination of methodologies, mindset, and tactics that align with your company's goals and target market. A successful sales strategy involves understanding your market, setting clear goals, aligning your sales process, and thinking beyond the immediate sale to nurture long-lasting customer relationships. Adaptability, continuous improvement, and a customer-centric mindset are vital to achieving consistent success in sales.

Here are some critical components of developing a successful sales strategy:

Understand Your Market and Customers:
- ❖ Conduct market research to identify your target audience, pain points, needs, and preferences.
- ❖ Create buyer personas to gain a deeper understanding of your ideal customers.

Set Clear Objectives:
- ❖ Define specific, measurable, achievable, relevant, and time-bound (SMART) goals for your sales team. These goals should align with your overall business objectives.

Segmentation and Targeting:
- ❖ Segment your market based on various criteria, such as demographics, geography, industry, or behavior.
- ❖ Tailor your sales approach to each segment, addressing their unique needs and challenges.

Competitor Analysis:
- ❖ Identify your competitors and analyze their strengths and weaknesses.
- ❖ Develop strategies to differentiate your product or service and highlight competitive advantages.

Sales Process Mapping:

- ❖ Create a well-defined sales process that outlines the steps from lead generation to closing the deal.
- ❖ Ensure your team understands and follows this process consistently.

Lead Generation and Prospecting:
- ❖ Develop a system for generating and qualifying leads.
- ❖ Implement various channels, such as inbound marketing, outbound calling, email campaigns, and networking, to reach potential customers.

Sales Enablement:
- ❖ Equip yourself with the right tools, resources, and training to effectively engage with prospects and close deals.
- ❖ Seek out ongoing support and coaching to improve your skills.

Sales Funnel Management:
- ❖ Monitor and track leads as you progress through the sales funnel.
- ❖ Identify bottlenecks and optimize the process for better conversion rates.

Value Proposition and Messaging:
- ❖ Craft a compelling value proposition that communicates how your product or service solves your customers' problems or meets their needs.
- ❖ Develop clear and persuasive messaging for different stages of the sales cycle.

Technology Integration:
- ❖ Utilize CRM (Customer Relationship Management) software and other sales technology to streamline processes, track interactions, and analyze data.
- ❖ Leverage automation to improve efficiency in tasks like lead nurturing and follow-ups.

Metrics and Analytics:
- Establish key performance indicators (KPIs) to measure the success of your sales strategy.
- Regularly analyze data to identify trends and areas for improvement.

Adaptability and Continuous Improvement:
- Stay flexible and be willing to adapt your strategy based on changing market conditions and customer feedback.
- Continuously learn from both successes and failures to refine your approach.

The Sales Process is a System

The sales process is a systematic approach that organizations and sales professionals use to convert leads into paying customers. It is a well-defined series of steps designed to efficiently and effectively guide potential customers through decision-making. When implemented correctly, the sales process can significantly increase business sales and revenue. Here's an overview of the sales process and how it serves as a system to generate more sales:

- **Prospecting:** The sales process begins with prospecting, where salespeople identify potential leads or customers who may have an interest in their products or services. Prospecting methods include cold calling, email marketing, social media outreach, and networking. The goal is to create a pool of potential buyers.
- **Qualification:** Not every lead is a good fit for the product or service being offered. In the qualification stage, sales professionals assess whether the lead meets specific criteria, such as budget, need, authority to purchase, and timeframe. This step helps prioritize leads and focus efforts on those most likely to convert.
- **Needs Analysis:** Once a lead is qualified, the salesperson conducts a needs analysis to understand the prospect's specific requirements and pain points. This involves asking questions and actively listening to the prospect to uncover their challenges and goals.
- **Presentation:** Armed with a deep understanding of the prospect's needs, the salesperson creates a tailored presentation demonstrating how their product or service can solve the prospect's problems and deliver value. This stage often involves showcasing features, benefits, and case studies.
- **Objection Handling:** Prospects may raise objections or concerns during the presentation. Skilled sales professionals effectively anticipate and address these objections by providing solutions and building trust. Handling objections is critical to moving the prospect closer to a purchase decision.
- **Closing the Sale:** The ultimate goal of the sales process is to close the sale. This stage involves asking for the prospect's commitment to move forward with the purchase. Salespeople may use various closing techniques to encourage prospects to make a decision.
- **Follow-up:** Even after closing the sale, the sales process continues with post-sale activities. This includes ensuring customer satisfaction, addressing post-purchase concerns, and potentially upselling or cross-selling additional products or services.
- **Referral and Relationship Building:** Satisfied customers can become advocates for your brand. Building and maintaining solid customer

relationships can lead to referrals and repeat business, thus extending the sales process beyond the initial transaction.

The sales process is not merely a set of arbitrary steps but a systematic approach to generating more sales and fostering long-term customer relationships. When implemented and continually refined, it becomes a powerful tool for businesses to drive revenue and achieve sustainable growth. However, you cannot always be locked in a systematic basic framework in the complex sales process. You have to be adaptable to "**Get Better at Sales**." The sales process is merely a roadmap for direction.

Personalized Sales Process

Having a well-defined and systematic approach is undeniably valuable. However, not all sales situations are created equal. A rigid, one-size-fits-all sales process can fall short in complex markets with diverse customer bases. This is where the art of adaptation comes into play.

Understanding the Complexity: Complex sales scenarios involve a multitude of factors, including:

- **Diverse Customer Profiles:** Your customers may vary widely in industry, size, needs, and pain points.
- **Longer Sales Cycles:** Complex sales often take longer and involve multiple decision-makers.
- **Unique Challenges:** Different markets present unique challenges and competitive landscapes.

The Need for Flexibility: In complex sales, adaptability is crucial. Here's why:

- **Tailoring Solutions:** Your customers have distinct needs. A one-size-fits-all approach won't address these unique requirements effectively. Adapting your sales process allows you to tailor solutions to individual clients.
- **Building Trust:** Trust is the cornerstone of complex sales. Adapting your approach shows that you genuinely listen and respond to your customers' concerns, strengthening the client-salesperson relationship.
- **Navigating Complex Decision-Making:** Complex sales often involve numerous stakeholders. Adapting means being flexible in your communication and influencing strategies to resonate with each decision-maker.

Expanding the Sales Process: Expanding your sales process involves going beyond the predefined steps and incorporating additional strategies:

- **Deep Market Research:** Invest time in understanding the intricacies of your target market. What are the current trends? Who are the major players? What are the pain points specific to this market?
- **Customized Content:** Create content that speaks directly to your customer segments within the complex market. Tailored messaging and resources can make a significant difference.
- **Relationship Building:** Complex sales often require longer-term relationships. Prioritize relationship-building activities, such as regular check-ins and personalized follow-ups.
- **Flexibility in Pricing:** Be open to flexible pricing structures, as different customers may have unique budget constraints or expectations.

The Benefits of Adaptation: By adapting and expanding your sales process, you can expect:

- **Increased Sales Success:** Tailoring your approach increases the likelihood of closing deals because you address customer-specific pain points and needs.
- **Enhanced Customer Loyalty:** Flexibility and adaptability build trust and foster long-term customer relationships.
- **Competitive Advantage:** Businesses that can pivot and cater to specific customer requirements gain a competitive edge in complex markets.

While a systematic sales process provides a solid foundation, it's essential to recognize that the real world of complex sales is dynamic and diverse. Adaptation and expansion are valuable and often necessary to thrive in such environments. Embrace flexibility, and watch your sales efforts flourish in even the most intricate markets.

Pro Tips for Developing a Personalized Sales Process

- Understand Your Customer: Start by creating detailed customer personas. Know their pain points, needs, and preferences.
- Segment Your Leads: Categorize your leads based on demographics, behavior, and past interactions.
- Tailor Your Messaging: Craft personalized messages and content that resonate with each segment. Address their specific pain points.

- Use Data and Analytics: Leverage customer data to make informed decisions and adapt your sales approach.
- Customized Sales Pitches: Create sales pitches and presentations that speak directly to each lead's unique needs and concerns.
- Adapt Communication Channels: Communicate with leads through their preferred channels, whether email, phone, or social media.
- Automate Repetitive Tasks: Automate routine tasks like sending follow-up emails, allowing sales teams to focus on more personalized interactions.
- Continuous Feedback: Collect feedback to refine and improve the personalized sales process.

Sustainability

When tying your sales strategy into the sales process and thinking a couple of steps ahead, focusing on the long-term relationship with your customers is essential.

Instead of solely concentrating on closing the immediate sale, consider the following:

- **Upselling and Cross-selling:** Identify opportunities to offer additional products or services that complement what the customer has already purchased.
- **Customer Success:** Ensure customers are satisfied with their purchases and provide ongoing support. Happy customers are more likely to become loyal advocates.
- **Anticipating Needs:** Stay attuned to your customers' evolving needs and proactively offer solutions before they even realize they need them.
- **Lifetime Value:** Consider the lifetime value of a customer rather than just the immediate revenue. A satisfied customer can bring in significant revenue over the long term.

The Pareto Principle

The Pareto Principle, with its simple yet profound insight that 20% of your efforts yield 80% of your results, is a game-changer in performance and time management. Embracing this principle empowers you to channel your energy and resources into the most impactful areas of your life, whether business, personal growth, or daily productivity. By focusing on the vital few instead of the trivial many, you unlock a newfound efficiency that allows you to accomplish more with less effort. It's not just a rule; it's a strategy that can revolutionize your work, giving you the leverage to achieve your goals and maximize your potential. Embrace the 80/20 mindset,

and watch how it transforms your performance, propelling you toward unparalleled success.

The Pareto Principle was discovered by an Italian economist named Vilfredo Pareto in the late 19th century. Pareto made this observation while studying wealth distribution in Italy in 1896. He noticed that approximately 20% of the population owned about 80% of the land. This led him to formulate the principle that came to bear his name.

The Pareto Principle has since been applied to various fields, including economics, business, management, and personal development. In business, it's often observed that roughly 20% of customers contribute 80% of profits. Identifying and focusing on these high-value customers can be a strategic move. In personal productivity, the principle suggests that 20% of your efforts will yield 80% of the results. So, prioritize tasks that fall into the 20% category to maximize productivity. In manufacturing or service industries, it's often found that 20% of defects cause 80% of problems. Identifying and addressing these critical issues can significantly improve quality.

The Pareto Principle has been empirically observed in various contexts, providing evidence of its validity. The Pareto Principle is more of a guideline than a hard and fast rule. It can be flexible. The exact percentages may sometimes be less than 80/20; they can vary. The principle is more about the unequal distribution of inputs and outputs than the specific 80/20 ratio. For instance, 20% of your customers contribute 80% of your sales, or 30/70, 10/90, etc. The key is recognizing that a small portion of your efforts or resources tends to generate a significant amount of your results.

Applying the Pareto Principle

Now that you know the Pareto Principle, let's consider how. It's easy to say focus 80% of your time on 20% of your customers. Figuring out where the Pareto Principle fits into your own life and determining how to best use your 20% effort for maximum impact involves a systematic approach of analysis and reflection.

Applying the Pareto Principle to your sales process can be highly beneficial. Identify the top 20% of your customers responsible for 80% of your sales. I will guide you to where to spend your time. Then, find the same type of customers in the top 20%. These prospects should be on your ideal target customer list. Focus your efforts on nurturing and retaining these high-value clients. Concentrating on the

most promising 20% of leads will likely convert into customers. This can save time and resources. What are your customers buying? Analyze your product or service offerings. Determine which products or services generate the most revenue and prioritize marketing and development efforts accordingly. You don't want to waste 80% of your time focusing on a 20% chance to sell it! You can apply the principle to time management by focusing on the 20% of activities that generate 80% of your sales results.

Strategies to Implement 80-20

First, start by clarifying your short-term and long-term goals. What do you want to achieve in your personal and professional life? Make a list of your priorities, both big and small.

- **Gather Data and Analyze:** Collect data and information related to your goals and activities. This could include your work tasks, personal projects, or any area you want to improve. Look for patterns and trends to understand which efforts contribute most to your desired outcomes.
- **Identify the Vital Few:** Analyze the data to pinpoint the 20% of activities or factors responsible for 80% of your results. These could be specific tasks, customers, products, or habits that disproportionately impact your success.
- **Eliminate or Delegate the Trivial Many:** Once you've identified the vital few, consider how to reduce or delegate the less impactful activities that fall into the remaining 80%. If possible, eliminate tasks that don't align with your goals or can be automated.
- **Time Management and Resource Allocation:** Allocate more of your time, energy, and resources to the vital few. Prioritize these activities in your daily and weekly schedule. Create a plan that ensures you consistently focus on what matters most.
- **Regular Evaluation and Adjustment:** Continuously monitor your progress and adjust your approach as needed. The Pareto Principle is not static; the 20% that matters may change over time as your goals and circumstances evolve.
- **Seek Feedback and Guidance:** Don't hesitate to seek feedback from mentors, colleagues, or trusted advisors. They can provide valuable insights into where your efforts are most effective and where improvements can be made.
- **Mindfulness and Reflection:** Regularly reflect on your actions and their impact on your goals. Mindfulness can help you stay aware of your choices

and ensure that you're consistently aligning your efforts with your priorities.
- ❖ **Continuous Learning:** Stay open to learning and improvement. Explore new techniques, tools, and strategies that can help you further optimize your efforts.
- ❖ **Celebrate Successes:** Acknowledge and celebrate your achievements resulting from your focused efforts. This positive reinforcement can motivate you to maintain the 80/20 balance.

Remember that the Pareto Principle is a flexible guideline rather than a rigid rule, so adapt it to your specific circumstances and objectives. Over time, as you become more attuned to where your 20% should be applied, you'll find yourself making more efficient and effective choices in various aspects of your life.

Make it Possible:
30 Proactive Actions Principles

Proactive actions are the soft skills needed to build the foundation for a successful life. These action principles are the behind-the-scenes skills you will use when using the hard skills of prospecting, closing, negotiating, building relationships, etc. You don't have to be a child to start on the right path for your behavior investments to start paying dividends. It helps, but it's better late than never!

"The best time to plant a tree was twenty years ago. The second-best time is now." – Chinese Proverb

Everyday actions you pay attention to and deliberately do will create the necessary habits that will inevitably lead to success. These are actions to live by that will start paying off for you right away and continue to pay dividends in the future.

1. **Make the Choice**

    First is making the choice. Decide what you want to do. Decide if you want to be successful. Consider your options, decide, and stick to it. Of course, think before you act because choices have consequences. You want to be sound in your choice but be comfortable knowing that your decision is yours and that you believe it to be the right one. The point is, don't pine over the choice between A and B. Make a choice quickly and move on it.

    Life is all about choices, and you are faced with choices every day, but most are inconsequential. *"Dang, I should have gotten in the right lane!" "Should I have one more cookie?"* You can be successful or unsuccessful, fulfilled or unfulfilled, happy or unhappy, likable or unlikeable, nice or rude, and so on. Nobody can act for you, and how you respond is all your fault.

"Successful people make decisions quickly and firmly; unsuccessful people make decisions slowly, and they change them often." –Napoleon Hill

2. Have a Positive Mindset

Being upbeat and positive will always be better than being downtrodden and negative. Being optimistic about any endeavor that you undertake will breed success. Throw away fears and get started. What's the worst thing that could happen? I am not saying I don't think about the bad things that can go wrong or ignore the burning buildings around you. Understand the negative and beware of the problems so you will know how to deal with them. Most things improve if you stay positive and stick to a plan, have a plan, and have perseverance.

"A negative mind will never give you a positive life." –Ziad K. Abdelnour

3. Set Goals

Have set goals and a solid plan to achieve those goals—set milestones for the goal to keep you on track. With a clear direction and solid plan to get there, all you need to do is do it!

"Setting goals is the first step in turning the invisible into the visible." –Tony Robbins

4. Start with the End in Mind

What does success look like? Envision your future and work backward. Great writers like Stephen King write the end of the story first. Knowing how the story ends, you can start with how to get there. Your life is very real but fictional in a sense. It hasn't happened yet; you can make it up as you go. Write your life story, but first, tell me how it ends. Write out the plan on how you will achieve your goal.

"To begin with the end in mind means to start with a clear understanding of your destination" –Stephen Covey

5. First Things First

Once you have worked yourself backward, start with the first thing you must do. Looking at your life as a whole or a big project, break it up into parts. What needs to be done first, then next? Putting one foot in front of the other and completing each required task before you take on the next one makes the project or task look possible. It's just small parts of the whole until, before you

know it, it's completed. Look at the task at hand and don't think about the three hundred other steps. Just think about the one you are doing. What's next? Then what is next?

"There is only one way to eat an elephant: a bite at a time." –Desmond Tutu

6. **Write Down a Plan**

   How will you do it after you set goals and see yourself as successful in the future? Consider what comes first. If you write down a plan, the plan becomes real. A plan sets up systems that come together to form a plan. Systems are the mechanisms that help you execute a plan of action.

   "If you fail to plan, you are planning to fail." –Benjamin Franklin

7. **Practice and Drill**

   I don't know everything, of course, but I do know that if you don't practice something, you won't get any better at it. You may never be the best, even with practice, but you can improve only with practice. You also will need to know if what you are practicing is right. How would you know if you are working on a skill but have nothing to compare it to? You may be practicing and getting it perfectly wrong.

   "Success has to do with deliberate practice. Practice must be focused, determined, and in an environment where there is feedback." –Malcolm Gladwell

8. **Set up a Routine**

   It's easy to get spread thin. With no structure, you might go crazy. Write out your priorities and set up a schedule to get things done. Routine is another term for the system. Start with the most critical tasks with the most weight first. Map out your day, week, month. Get up at the same time, work out at the same time, make time for reading at the same time, and drink coffee at the same time. Setting up a routine will simplify your life and reduce stress. Being predictable does not mean you are boring; it just means you are predictable. After that, get creative and add things to your routine that you don't have now. Working out, reading, family time, anything else? Even if you have a routine for the first two hours of your day, it will help you set the pace and start on the right track. Your routine will set you up for success.

   "If you think adventure is dangerous, try routine; it is lethal." –Paulo Coelho

9. **"KISS IT"**

    Keep It Super Simple! Don't overcomplicate your goals or the plan to get there. Identify the critical components of what needs to be done. Focus on priorities and make it happen. Don't let yourself enter a paralysis-by-analysis state where you look at every angle and cover every base. There are a lot of buts and what-ifs.

    "Keep it simple and focus on what matters. Don't let yourself be overwhelmed." –Confucius

10. **Go Above and Beyond**

    "You reap what you sow" is a proverb that says present actions inevitably shape future consequences. This shouldn't be confused with karma because karma is described as getting payback from the "universe," good or bad. The "You reap what you sow" is based on action, and you will get paid back by others who are appreciative and thankful for those actions. How can you do more than what's expected? Do something when you don't expect anything in return. Put out more than you receive, and someday, someone will notice. This builds character, habit, and trustworthiness. If you always do more than expected, people will rely on you to do what needs to be done. You will become indispensable. Take a personal initiative and do something extra. "That's not my job" or "Some else will do it, don't worry about" attitudes won't get you further in life but average.

    "Go the extra mile, it's never crowded." –Wayne Dyer

11. **Tell the Truth**

    People don't have time for lies or liars. If someone doesn't know that you are lying to them at first, they will find out. Promise. You can't move on from a lie. It affects your relationships and your mind. The lie will eat at you and compound, giving you a slight edge for failure. A lie will grow because your story must jive after another is told to cover the first one, then another after some more questions. Then you forget what you said because it didn't happen how you described, not how you remember. Lies lead to denial, and denial leads to mistrust. People can forgive a mistake and may forget about it, but they won't forget the lie.

    "If you tell the truth, you won't have to remember anything." –Mark Twain

12. **Stay Principled**

Don't sell your soul to the devil for anything. Stay true to yourself and live by your values. If you break your value principles, you break yourself down, becoming a person you may not like. I wouldn't say stubborn, hardheaded, or closed-minded, but stand by your beliefs. You need to be able to look in a mirror and see what you like. Others may not like your values or your willingness to stand up for what you believe is right, and you know what? That's okay. You don't need someone else's approval to love yourself. Keep your promises to yourself, and do not worry about what others think of you. Live your life on your terms, and don't be ashamed of who you are.

"Obey the principles without being bound by them." –Bruce Lee

13. Stay Focused

SQUIRREL! Do you feel like you are pulled in several different directions at once? It's okay to take on multiple projects or responsibilities, but it's not okay to be Mr. 95 percent. Being Mr. 95 percent is starting different projects and leaving them 95 percent completed because it's good enough to start or try to finish another, leaving all projects 5 percent incomplete. Prioritize which tasks are most essential to complete and stay on the task until it's complete before starting another task. This practice will translate into your personal life while spending time with people. Staying focused on the person you are with and being engaged is necessary to build relationships. There is plenty of time to complete everything; stay focused on completing what you are working on first.

"Wherever you are, be there. –Jim Rohn

14. Stay Inspired

Take your motivation up a level and be inspired. Who inspires you? Why do you get up every day? What are you working for? Motivation falls off the cliff daily. You need something more to keep the dream alive and to keep working toward something. You are always motivated, but what is the question. Having an overarching reason will keep you inspired. You are not inspired to eat a steak; you are motivated to because you are hungry. You aren't motivated to be financially independent and have a house in Malibu with ocean views and a gourmet kitchen with a twelve-burner stove, but you are inspired to. You are not inspired to get up and work every day in a career you hate; you are motivated to because you are inspired to see your kids off to college so they

can accomplish their goals and realize their dreams. Or maybe you are inspired by your work because you are making a difference and being the change you want to see. Being inspired will keep a smile on your face and keep you focused on what is most important to you other than your basic needs, no matter what your cause is.

"My mission in life is not merely to survive, but to thrive" –Maya Angelou

15. Keep Your Promises

First, look in the mirror and learn to keep promises to yourself and keep promises made to others. Stick with your decisions and work to achieve your goals. Find a way to keep promises. Work very hard to do so. Go the extra mile. Things that are out of your control come up, so if you need to break a promise, make it right to the people you promised. If you can't keep any promises, then don't make them. Be honest. If you break enough promises, you will break trust and won't be able to be counted on.

"To be responsible, keep your promises to others. To be successful, keep your promises to yourself." –Marie Forleo

16. Show Gratitude

Say thanks. Geez, showing gratitude is not difficult. Hold the door. Help someone. Don't be that bad guy; just be nice. You will be paid back for what you put out in the world. Take deliberate action. Take the time to thank the people who help you. Take the time to show appreciation to friends, family, and strangers. Send an email or text, call, buy dinner, or thank people in person. Let people know you are grateful.

"Tart words make no friends. A spoonful of honey will catch more flies than a gallon of vinegar." –Benjamin Franklin

17. Commit to Perseverance

In other words, don't give up. See things through to the end, and don't give up until you have exhausted your resources. You had a setback. Now what? Keep going. Stick with the plan. Execute, and don't give up until you have succeeded.

"Many of life's failures are people who did not realize how close they were to success when they gave up." –Thomas Edison

18. Have Patience

Everyone makes mistakes, but you, right? Put yourself in another person's shoes and slow down. This is patience in the moment with people. Sometimes, you need to wait for people or count on them. Remember gratitude. But don't sit around waiting for something to happen. Make things happen. Just be patient to know there are no get-rich-quick schemes, and things take time. Hope is not a strategy, and waiting isn't a plan. Have patience with people and yourself, but don't wait! Patience is essential because it is the only practical way compounding can occur in your habits, physical health, and financial health. These actions need time to build. Like a seed to a tree, improvements take time to grow. It takes time for a tree to grow from seed.

"Things may come to those who wait but only the things left by the people who hustle." –Abraham Lincoln

19. Be Loyal

Nobody wants a fair-weather friend. So why are you one? Maybe not, but the point is you need people in your life who will stick with you throughout all your ups and downs. You also need to be this person. Not with everyone, just with the people who deserve it back. Everyone is weird or even crazy. Who cares? If we were all the same, this place would get pretty dull.

"Nothing is more noble, nothing more venerable, than loyalty." –Cicero

20. Be Loving

Go out of your way and love. Show love to your friends, family, spouse, dog, and strangers. If you are in a committed relationship and want it to last, be loving—just love. When you are in a bad mood or get upset at someone for doing something you can't stand, reverse your attitude and show love. Change your perspective around and deflect your loving actions toward them.

"Love does not dominate; it cultivates." —Johann Wolfgang von Goethe

21. Help Others Succeed

If you help someone else succeed, you have a good idea of how to help yourself. Be in service to others. You can do it yourself once you have mastered making money for others and helping them become rich. Consider it practice. Use the experience and any money you get from helping others and then apply that knowledge to help yourself.

"If I then, your Lord and Teacher, have washed your feet, you also ought to wash one another's feet. For I have given you an example that you should do just as I have done to you." John13:1-15 ESV

22. Keep the Faith

Faith deals with more than just religion. Having faith is seeing success even when you haven't achieved it yet. Faith is knowing that you have the ability and will all work out. Know that you will succeed, even though you may not know how to get there or what lies ahead. Just know that you can do it. You have the end in mind and trust that it will work. Faith is a component of success, not the driver.

"Faith is to believe what you do not see: the reward of this faith is to see what you believe." –Saint Augustine

23. Practice Positive Self-Talk

Telling yourself you can do it and working through the plan promotes faith. If you tell yourself you can't do something, then as Henry Ford said, "You're right." Set your boundaries, make your own rules, and set your limitations. You must know your limitations, but you can't take them to the extreme and have every limitation. Being positive is far more helpful than being negative and doesn't hurt anybody.

"Whether you think you can or can't – you're right." –Henry Ford

24. Speak Up

Self-promotion is good right up until the point of bragging. You must let people know of your skills and what you can do for them throughout your life and in every capacity. Let people know your ideas, your goals, and your skills. If you don't speak up, you will go unnoticed and pass by for the next person, who may not be as good as you, whether it means taking your potential job, promotion, or future spouse. As the old saying applies, the squeaky wheel gets the grease. If you don't speak up, someone else will take advantage of what you could have called. The world is a competitive place. Don't let yourself get passed by. The only one looking out for you is you.

"Just remember always to be yourself and do not be afraid to speak your mind to others." –JA Redmerski

25. Investments in Yourself

A. **Mentally**

   It's impossible to have too much knowledge. However, it is possible to have too much knowledge of things that won't get you anywhere. What is important to you? Invest in what interests you and focus on how you can grow that knowledge. If you like woodworking, reading ten books on cooking doesn't make sense. Crave the knowledge that will intrigue you, think about things differently, and be well-rounded.

   "Invest in yourself; your knowledge is the engine of your wealth." – Warren Buffet

B. **Financially**

   Get in the habit of saving. Pay yourself first. Don't work your whole life and have nothing to show for it. Save at least 10 percent of each paycheck to start and ramp it up as you go. Don't procrastinate on this task. Start saving right away. After saving, start investing.

   "Don't save what is left after spending, but spend what is left after saving." –Warren Buffet

C. **Physically**

   Get to the gym. Walk. Run. Get on the bike. Play Tennis. Play basketball. Whatever. Make the time. Period. Your excuse not to do something physical is that—an excuse. It's your choice. Then get a workout buddy. Hold each other accountable.

   "Take care of your body; it's the only place you have to live." –Jim Rohn

26. Let Your Mom Worry

Seriously. Focus only on what you can control and deal with what you can't. Plus, your mom will worry about you enough for the both of you.

"I've had many worries in my life, most of which never happened." –Mark Twain.

27. Surround Yourself with People Better Than Yourself

I only pretend to know everything. I will happily admit I am smart enough to listen to people with something to say. I ask many questions to people with more ability than me, and I understand they have made mistakes I don't want to make. To surround yourself with better people, you need to be like those

people. Like attracts like; you may need to change who you are to be who you want to be. It's up to you to change and decide to be a better person.

"Every person you meet knows something you don't. Learn from them." – H. Jackson Brown, Jr.

28. Display Humility

Humble pie is my favorite dessert. Until their last days on the job, the greatest football player and greatest basketball player to play their respective sports, Tom Brady and Michael Jordan, knew their roles. They were players, and they listened to their coaches. They have a part of the team with a job to do. Their egos remained in the locker room, and they remained humble on the field. When you think you are more significant than the team or the best, you will let your team down and lose collectively. There will always be someone bigger, better, and more intelligent than you. Do the best that you can do, know your role, and leave your ego at the door.

"True humility is staying teachable, regardless of how much you already know." –Unknown

29. Elicit Feedback

Elicit feedback and listen to it! How will you know you're getting better? How will you know you've been good at something? Asking others for feedback and trying to improve is a quick way to grow. Even if you don't want feedback, you should still listen to what others tell you. For example, if you're walking into a grocery store and you hold the door open for a seasoned citizen, and they're very appreciative and smile and say, "Thank you," are you more likely to continue that behavior? Yes, a good idea would be to continue that behavior because you were rewarded for it. Other people will let you know if your actions are good or bad. How did I do? What can I do better? How can I change? These are great questions to ask to elicit feedback.

"We all need people who will give us feedback. That's how we improve." –Bill Gates

30. Execute

Get moving before rigor mortis sets in! All other actions can only be accomplished with the final proactive action, likely the most important one because it focuses on doing. Nothing will get done if you don't start. You can

write all the plans down, think about your approach, and rethink it repeatedly, but without execution, there is no accomplishment.

Setting goals is the easy part. Following through is the hard part. Focus on doing, not perfection. Decide and change your plans as you go, but get started. Throw fear out the window and start executing your strategy for success today.

Hope is not a plan. Having hope is waiting for someone else to do something for you: "I hope they call," or "I hope it works out," or "I hope someone buys something today." Having faith is what you do for yourself. Trust your ability to know that you will execute and accomplish things on your merit.

"A good plan, violently executed now, is better than a perfect plan tomorrow"
–General George S. Patton

Chapter 2
Know Thy Customer.

People have different personalities, and not everyone will think the same way as you. This diversity in thinking and preferences is where the Platinum Rule becomes instrumental. By treating others how they want to be treated, we cater to their unique communication styles, fostering better understanding and connection. This approach is critical when conveying your message to individuals with personality types different from your own. Knowing who you are talking to and how they prefer to be engaged is not just effective; it's essential in achieving successful communication.

The saying, "Treat others how they want to be treated," is most commonly attributed to Dr. Tony Alessandra. Dr. Alessandra is a well-known author, speaker, and expert in the fields of marketing and communication. He introduced this concept to emphasize the importance of understanding and respecting others' unique preferences and perspectives, particularly in the context of business and personal relationships.

The Platinum Rule shifts the focus from self to others. This concept contrasts with the Golden Rule and is often cited in discussions about effective communication and interpersonal skills. The Platinum Rule is a twist on the Golden Rule, which traditionally states, "Do unto others as you would have them do unto you." The Platinum Rule, on the other hand, advises, "Treat others the way they want to be treated."

It emphasizes the importance of understanding and respecting other people's feelings, opinions, and cultural backgrounds. This rule is particularly valuable in diverse environments and interpersonal relationships, as it promotes empathy, effective communication, and a deeper understanding of others' perspectives.

In practical terms, the Platinum Rule encourages us to put ourselves in others' shoes and consider their unique preferences and needs. This approach is especially relevant in professional settings such as customer service, management, and sales, where understanding and catering to individual preferences can significantly impact outcomes and relationships.

Adapting your communication style to match the personality type you are speaking with, as suggested by the DiSC personality tests, is a practical application of the Platinum Rule. This doesn't mean losing your authenticity; instead, it's about flexibly expressing different parts of your personality that resonate with the dominant aspects of another's personality.

Like Attracts Like

Being relatable, approachable, and agreeable is crucial in communication. Understanding personality types helps in finding common ground without losing your individuality. The goal isn't to mimic or mirror your customer in a way that feels insincere or 'creepy,' but to genuinely connect by finding shared values or perspectives. This approach reinforces the importance of being human and genuine in interactions.

Square Peg in a Round Hole

Envision a scenario where a square peg is forcefully jammed into a round hole. It's an image that vividly illustrates the wrong approach to relatability. Like trying to fit that square peg into a shape it doesn't naturally conform to, attempting a one-size-fits-all strategy with customers often leads to a lack of authenticity and missed connections. It's a misalignment that customers can sense, leaving them feeling unheard and unvalued.

The strategies employed to interact with and understand customers can significantly impact business success. This concept is often metaphorically described using the analogy of fitting a square peg into a round hole versus fitting a round peg into a round hole. Even if you can't quite fit perfectly, find a way to fit well enough to relate to your customer.

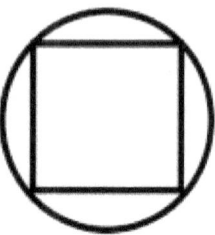

Square Peg Approach: This approach symbolizes rigidity and a one-size-fits-all mindset. It is like trying to fit a square peg into a round hole, which naturally doesn't align well. This method can be problematic in customer relations because it often fails to consider each customer's needs, preferences, or personality. When salespeople or businesses adopt this approach, it can appear inauthentic or insincere. Treating every customer the same without acknowledging their uniqueness may lead to missed opportunities for genuine connections. It can result in adverse outcomes like customer disengagement, dissatisfaction, or even loss. The metaphor suggests that forcing a fit, even if it's rough and imperfect, is not effective.

Round Peg Approach: In contrast, this approach is about flexibility and personalization. It compares fitting a round peg into a round hole, implying a natural, seamless alignment. This method focuses on adapting to the unique characteristics of each customer. It emphasizes the importance of genuine connections and acknowledging and respecting individuality. The round peg approach is customer-centric; it involves listening, understanding, and responding in a way that feels authentic to the customer. It's about placing the customer at the center of the strategy and tailoring interactions and offerings to suit their specific needs and preferences.

In sales, particularly in complex scenarios, applying the Platinum Rule is critical to understanding and guiding your target customer's buying process. Not all customers are familiar with making significant purchases, and as a sales professional, your role extends beyond selling – it involves guiding customers through the buying process. By aligning your communication and sales approach with the Platinum Rule, you not only sell but also enable your customers to make informed decisions in a way that feels comfortable and natural to them.

Customer Persona

Before diving into the buying process, it's essential to understand your target customer persona. A customer persona is a detailed representation of your ideal customer, encompassing their demographics, preferences, pain points, and behaviors. Creating a robust customer persona helps you tailor your sales approach to match the needs and preferences of your potential buyers.

Create a target customer persona that represents the segment of your target audience. This persona is a fictional character that embodies the characteristics, behaviors, and goals of your ideal customer. By humanizing your target audience, you can better empathize with their needs and challenges. This will help you visualize your ideal customer and promote focus and insight into whom you are looking for. Like a most wanted poster, any person can be a suspect until the picture of the person wanted is posted. Then, you can narrow it down to that single person. I understand people are unique, and you will sell to multiple people. But if you sell double-seat strollers, you would focus on women with young children and two children. This gives you focus. However, there are millions of families with two young children. You wouldn't focus on your ideal customer as an 18-year-old male.

> **Here are some areas of interest to cover when creating your ideal customer profile.**

- ❖ **Identify Customer Goals:** Understanding your target customer's goals is essential for aligning your products or services with their aspirations. Conduct surveys, interviews, and focus groups to gain insights into what your customers want to achieve. By uncovering their desires, you can tailor your offerings to provide the desired solutions.
- ❖ **Pinpoint Barriers to Success:** Customers will face barriers that impede their progress toward their goals. These barriers can be categorized into three main types:
    a. **Functional Barriers:** These obstacles arise from practical limitations such as lack of knowledge, limited resources, or insufficient time. Identify the functional barriers your customers encounter and explore how your products or services can alleviate them.
    b. **Emotional Barriers:** Emotional barriers stem from fear, uncertainty, or negative past experiences. It is crucial to understand your customers' emotions and address them effectively. By building trust and offering a positive experience, you can help customers overcome these emotional hurdles.

c. **Financial Barriers:** Financial constraints often hinder customers from achieving their goals. Evaluate your pricing strategies and consider offering flexible payment options or financial assistance to mitigate this obstacle.

- **Effective Marketing Strategies:** Crafting marketing messages that resonate with your target audience is vital for attracting and retaining customers. Use the language, tone, and channels that align with their preferences. Use social media, content marketing, and personalized advertising to reach your target customers effectively.
- **Continuous Feedback and Iteration:** Customer preferences and barriers evolve. Stay attuned to their changing needs through regular feedback mechanisms such as surveys, reviews, and customer support interactions. Continuously refine your offerings and marketing strategies to stay relevant and meet the evolving demands of your target audience.
- **Demographics:** Understanding the demographic information of your ideal customer is crucial. This includes factors such as age, gender, location, occupation, income level, and education. This information helps you create a clear picture of your target audience and tailor your messaging and marketing strategies accordingly.
- **Needs and Goals:** To relate to your ideal customer, you must understand their needs and goals. What challenges do they face? What are they trying to achieve? By identifying their aspirations and pain points, you can position your products or services as solutions that directly address their specific desires, making a solid connection with them.
- **Preferences and Interests:** Knowing your ideal customer's preferences and interests helps you create relevant, engaging content that resonates with them. Are they interested in specific hobbies, sports, or entertainment? What are their preferred communication channels? This information enables you to tailor your marketing efforts to align with their preferences and effectively reach and engage them.
- **Behavior and Buying Patterns:** Understanding how your ideal customers behave and make purchasing decisions is vital. What influences their buying choices? Do they conduct thorough research or make impulsive decisions? By understanding their buying patterns, you can optimize your sales strategies and create a seamless customer journey that caters to their preferences, increasing the likelihood of conversion.
- **Pain Points and Objections:** Identifying your ideal customer's pain points and objections is critical to building a solid relationship with them. What barriers do they face when considering your product or service? What objections might they have? By addressing their concerns proactively, you can overcome objections and position yourself as a trusted partner who understands and empathizes with their challenges.

Defining your target customers and understanding their barriers are fundamental to building a successful business. Implementing effective marketing strategies and maintaining a feedback loop will ensure your offerings remain relevant and aligned with your target audience's evolving requirements. By focusing on your customer's success, you will foster long-term relationships and drive the growth of your business.

Personality Questions

To identify who you are speaking with, you must ask prospects questions to determine their personality type. These question frameworks must be modified and tailored to your situation.

We will reference the DiSC personality profiles to identify a person's personality type. The DISC personality assessment, developed by psychologist William Marston in the 1920s, uses a four-quadrant model to describe personality traits. The name "DISC" is an acronym representing the four primary personality types: Dominance, Influence, Steadiness, and Conscientiousness. You could devise alternative names for these categories if you want to rename the DISC personality types.

> Here's a brief description of each of the four DISC personality types:

Dominance (D):

- ❖ Dominance-oriented individuals are often described as assertive, results-driven, and competitive.
- ❖ They tend to be goal-focused, enjoy taking charge of situations, and are willing to make tough decisions.
- ❖ Dominant individuals are typically direct and may be assertive or even forceful in their communication style.

Influence (I):

- ❖ Influence-oriented individuals are often seen as outgoing, sociable, and enthusiastic.
- ❖ They are natural communicators and thrive in social settings, building relationships quickly.
- ❖ People with an influencing personality are often optimistic and persuasive, motivating others with energy and enthusiasm.

Steadiness (S):

- ❖ Steadiness-oriented individuals are known for their patience, empathy, and calm demeanor.
- ❖ They value stability and teamwork, preferring a harmonious and supportive environment.
- ❖ People with a steadiness personality are often excellent listeners and dependable team members.

Conscientiousness (C):

- ❖ Conscientiousness-oriented individuals are detail-oriented, organized, and analytical.
- ❖ They prioritize accuracy, structure, and quality in their work and decision-making.
- ❖ People with a conscientious personality are often seen as systematic and cautious, focusing on precision and thoroughness.

Most people exhibit a combination of these personality traits to varying degrees, and the DISC assessment typically provides a profile that highlights one's dominant characteristics. Understanding your DISC profile and those of others can be valuable for improving communication, teamwork, and interpersonal relationships.

Uncover Their Personality

You can use these questions as a guide and work them into conversations with prospects to uncover dominant personality traits and tailor your messaging and process to accommodate them. Knowing what questions to ask is half the battle, and you can ask questions and not be able to analyze the answers.

Here are some of the solutions that may look from prospects on what their personality type is more dominant.

Dominance (D):

Question: How do you usually tackle the exciting challenges that come your way in your line of work?
Example Answer: *"I dive right in and take charge. I'm all about finding solutions and getting things done efficiently."*

Question: What motivates you to take charge and steer your team towards success?
Example Answer: "I thrive on competition and driving results. Leading others to achieve goals is where I feel most in my element."

Question: Could you share a story about a time you had to make a quick decision to move things forward?
Example Answer: "Sure, there was this project where time was running out. I analyzed the situation, made a bold call, and it paid off big time."

Influence(I):

Question: Tell me, what kinds of social activities or interactions energize you the most outside of work?
Example Answer: "I love attending parties, networking events, and meeting new people. Socializing is like fuel for me."

Question: What factors influence your motivation when considering your ideal work environment?
Example Answer: "A vibrant and collaborative workspace where I can interact with others. I excel when I can connect and inspire."

Question: Can you describe an instance where your ability to connect with people was vital in achieving a common goal?
Example Answer: "Absolutely, there was a time when I rallied the team for a challenging project. My ability to influence others' perspectives was crucial."

Steadiness(S):

Question: How do you usually adapt when things unexpectedly change in your day-to-day routine?
Example Answer: "I tend to stay calm and find ways to make the best of the situation. I value stability and consistency."

Question: As you reflect on your teamwork experiences, what roles resonated most with you and why?
Example Answer: "I often find myself in supportive roles, ensuring everyone's on the same page and creating a harmonious atmosphere."

Question: Could you recall a moment when your ability to bring a sense of calm positively impacted a stressful situation?
Example Answer: *"Certainly, during a tense meeting, I stepped in to mediate and ease tensions, which helped the team work together more smoothly."*

Conscientiousness (C):

Question: When it comes to tasks that demand attention to detail and meticulous planning, how do you usually approach them?
Example Answer: *"I meticulously plan and analyze every step. Precision and accuracy are essential to me."*

Question: What elements of organization matter the most to you in your work or personal space?
Example Answer: *"I'm all about order and structure. My workspace is organized down to the last detail, and I value the same in my work."*

Question: Can you share an example of when your dedication to precision and accuracy paid off?
Example Answer: *"Certainly, I once caught a critical error in a report just before it went out. My thoroughness saved the day."*

These example answers should show how prospects' responses can align with different personality types. Remember that people may exhibit a mix of traits; the overall pattern can help you determine their dominant personality tendencies. You don't want to limit yourself to asking open-ended questions and listening to their answers.

Here are some strategies you can employ to assess the personality type of a prospective client quickly:

- ❖ **Observe Communication Styles:** Pay close attention to how the prospective client communicates. Do they focus on results and outcomes (Dominance)? Are they friendly and sociable (Influence)? Do they seek harmony and stability (Steadiness)? Are they detail-oriented and systematic (Conscientiousness)? Their communication style can provide valuable clues about their dominant traits.
- ❖ **Listen Actively:** Listen attentively to their responses and language. Do they use assertive and direct language (Dominance)? Are they enthusiastic and expressive (Influence)? Do they focus on maintaining relationships and creating a supportive environment (Steadiness)? Do they emphasize accuracy and details (Conscientiousness)?

- ❖ **Observe Body Language:** Nonverbal cues such as body language, facial expressions, and gestures can reveal much about a person's personality. For instance, a person with a dominant personality might have a confident posture, while someone with a conscientious personality might pay close attention to details in their environment.
- ❖ **Pay Attention to Interests:** Inquire about their hobbies, interests, and how they spend their free time. People often gravitate towards activities that align with their dominant traits. For example, someone with an influential personality might enjoy social gatherings and events.
- ❖ **Notice Decision-Making:** Observe how they make decisions. Do they make quick, assertive choices (Dominance)? Do they consult with others and seek opinions (Influence)? Do they take time to weigh options (Conscientiousness) carefully? Do they prioritize stability and harmony in their decisions (Steadiness)?

The goal is to gather enough information through these strategies to make an educated guess about their dominant personality traits. Remember that individuals are complex and may exhibit traits from multiple personality types, so be open to various possibilities. Once you know who you are dealing with, you can ask more pointed questions they will feel more comfortable answering.

Chapter 3
Get Better Meetings.

The way you can get sales faster is to have more face-to-face meetings. And who to have a meeting with can also be a consideration. Meeting with someone you have never met can be a challenge. Let's look at some strategies to help you get more and better meetings.

*There's a catch. These meetings need to have a purpose.

Face-to-Face meetings

> "I don't want to be just a voice on the phone. I must get to know these guys face-to-face and develop a sincere relationship. That way, if we run into problems in a deal, it doesn't get adversarial. We trust each other and have the confidence we can work things out."
>
> — Wayne Huizenga

Sales interactions work best face-to-face. The following article referenced proves this. Getting on the same wavelength with people and understanding is done face to face. You will get more sales faster if you see people face to face. You are better off seeing someone in person to get in sync with them and make a connection.

Face-to-face meetings are essential. A connection can be greater on a subconscious level. Thoughts are things you can pick up on invisible wavelengths sent by people you speak with. Connecting on a deeper level goes beyond verbal communication and body language.

"Brain Waves Synchronize when People Interact" July 1, 2023, *Scientific American* discusses the emerging field of collective neuroscience, where researchers study how brain waves synchronize during social interactions.

The article highlights that when people engage in conversations or share experiences, their brain activity becomes synchronized. This synchronization is observed in corresponding areas of the brain and is akin to dancers moving in harmony. Auditory and visual brain regions show similar responses to stimuli, while higher-order brain areas exhibit synchrony during complex tasks like understanding and interpreting information. This phenomenon of being "on the same wavelength" is natural and visible in brain activity.

The author of the article, Lydia Denworth, describes her participation in a study where she and another participant engaged in joint storytelling while their brain activity was measured. The study sought to uncover both brainwave synchrony and how brains align at a deeper level of understanding. The researchers found correlations in specific brain regions while participants told joint stories, especially in areas related to memory and narrative construction.

The research indicates that brainwave synchronization has implications for learning, enjoyment, relationships, and social evolution. For instance, in classroom settings, students' brain patterns align with their teacher's, potentially enhancing learning. Musical performers and their audience also exhibit synchronized brain activity, suggesting a link between enjoyment and synchrony.

Close relationships, such as romantic partners and close friends, show higher degrees of brainwave synchrony compared to more distant acquaintances. The article goes deep into the mechanisms behind brainwave synchrony, suggesting it goes beyond shared experiences and indicates something more profound. Researchers aim to understand the exact nature of synchrony by analyzing brain activity during interactions. Studies involving functional magnetic resonance imaging (fMRI), functional near-infrared spectroscopy (fNIRS), and electroencephalography (EEG) provide insights into brain synchrony across different species.

Face-to-Face Meetings and Connection

The concept of brainwave synchronization discussed in the article can indeed correlate with the effectiveness of face-to-face meetings in sales interactions as opposed to over-the-phone conversations. When individuals engage in face-to-face interactions, a higher degree of brainwave synchronization often occurs due to the presence of visual cues, body language, and nonverbal signals. These

elements are essential for conveying emotions, intentions, and subtleties that are challenging to convey solely over the phone.

Face-to-face meetings offer a more comprehensive form of communication, encompassing spoken words, facial expressions, gestures, and posture. This multifaceted mode of communication contributes to fostering a stronger sense of connection and understanding between salespeople and potential clients. Building rapport and establishing a genuine connection is generally easier in person, and the synchrony of brain activity during face-to-face interactions can contribute to building trust and rapport, which are pivotal in the sales process.

Another advantage of face-to-face meetings is the ability to receive instant feedback through visual and auditory cues. Salespeople can adapt their approach in real time based on the client's reactions and responses, leading to a more dynamic and responsive conversation. Additionally, emotional cues are more pronounced in face-to-face interactions, and emotions play a significant role in decision-making.

Sales interactions that elicit positive emotions and resonate with clients' needs are more likely to yield positive outcomes. The personalization of the sales pitch is also enhanced during in-person meetings. Salespeople can tailor their approach based on the client's visual cues and subtle signals, resulting in a more tailored and effective presentation. Face-to-face meetings facilitate a combination of verbal and nonverbal cues, leading to better information retention and a clearer understanding of the product or service being discussed.

Solutions Provider – Problem Solver

Know one problem you can solve and then find the prospects with that problem. The problem with that. This can be like finding a needle in a haystack. Everything perfect needs to happen. So, the alternative is to find people unaware they have a problem and then help them understand that there are better ways to do business. Once you get people to see they have a problem, you create a needle in a stack of needles.

In your industry, you have customers with similar problems to each other. There are different situations and levels of awareness, but the point is that you or your company has figured out the gap in the market, and you are there to fill it. Your product or service exists specifically to solve a particular problem for a specific market vertical. There are opportunities even when they are not apparent.

Where to start?

- Put yourself in a position to meet the right people.
- Go where they are.
- Of course, you know this. But are you doing it? Implementation is the prime word.

It's knowing what to say.

- Have scripts or rehearsing and practicing what to say.
  - The biggest holdup to starting a conversation is not knowing how to start.
- How can you pique interest?
  - Keep it simple. The best message is delivered at a third-grade level.

Define your why.

- Why does your customer benefit?
- Why do you help people?
- Why does someone need you?
- Why are you offering the solution?

We give people too much credit and expect them to connect the dots for themselves. This rarely happens. You spent the last who knows how long and spoke to countless people about the product and company that you "get." It took months, if not years, to sell yourself on what you sell. Now, you expect people to understand that they will get what you do within the first 30 seconds of meeting them.

Success hinges on your ability to connect with potential clients, identify their needs, and present solutions that truly resonate. Yet, this process often feels like finding a needle in a haystack, especially when looking for prospects with specific problems to solve. But what if there was an alternative approach, a way to find those who might not even realize they have a problem and guide them toward a better way of doing business?

Write Down Your Problem-Solving Statements.

Writing a problem statement is essential in various contexts, from research projects to business initiatives. A well-crafted problem statement helps you clearly define the issue you're addressing and sets the stage for finding solutions.

Different situations call for different statements depending on where you are in the process. Take a minute to write down some statements that will encapsulate your solution and get your customers' attention.

- ❖ One sentence to get their attention.
- ❖ One paragraph tells them more to set the meeting.
- ❖ One page tells the whole story during the meeting or presentation.

Position your message to set up the meeting based on how you can help or what you do.

Here's a guide on how to write effective problem statements:

- ❖ **Identify the Problem:** Start by clearly identifying the problem you want to address. Be specific and concise. Avoid vague or broad statements. Consider the problem, where it occurs, who it affects, and why it's significant.
- ❖ **Define the Scope**: Determine the boundaries of the problem. What aspects are within the scope of this problem statement, and what aspects are not? Defining the scope helps prevent the problem statement from becoming too broad.
- ❖ **Provide Context:** Explain the context or background of the problem. What led to the emergence of this issue? Include relevant data, statistics, or trends illustrating the problem's significance.
- ❖ **State the Impact:** Describe the impact or consequences of the problem. How does it affect individuals, organizations, or the community? Clearly articulate the adverse outcomes or risks associated with the situation.
- ❖ **Highlight the Need for a Solution:** Emphasize why addressing this problem is essential. What benefits or improvements will come from finding a solution? Convey the urgency or relevance of solving this problem.
- ❖ **Avoid Assumptions:** Ensure your problem statement is based on facts and evidence, not assumptions or personal opinions. Back up your claims with data or research.
- ❖ **Use Clear and Concise Language:** Write the problem statement using clear and straightforward language. Avoid jargon or technical terms that may confuse the reader.
- ❖ **Make it Actionable:** Ideally, a problem statement should lead to actionable solutions. It should inspire further investigation, research, or problem-solving efforts.

- ❖ **Review and Revise:** After drafting your problem statement, review it carefully. Make sure it accurately represents the problem and is free from ambiguity. Revise as needed to improve clarity and precision.
- ❖ **Seek Feedback:** Get feedback from colleagues, experts, or stakeholders familiar with the issue. They can provide valuable insights and help refine the problem statement.

Chapter 4
Find a Needle in a Needle Stack.

The best way to get sales faster is to stack the deck in your favor. Instead of looking for something that might be impossible to find, you need to be in the right place at the right time every time for this to work. Instead, create the environment that you need to be successful.

Set Yourself Up for Success

"Stacking the deck in your favor "means taking proactive steps to create a favorable environment for sales success. This involves strategically targeting and qualifying leads, leveraging networks and relationships, and continuously adapting your approach based on market insights. By focusing on these areas, you can increase your sales efficiency and effectiveness rather than leaving it to chance.

Traditionally, sales success is often attributed to being in the right place at the right time – essentially, a matter of luck. However, consistently relying on chance is not a sustainable strategy. Instead, the focus should be on creating opportunities rather than waiting for them.

This means actively constructing an environment or circumstances conducive to sales success. This could involve positioning yourself in networks or platforms where your potential customers are likely to be or tailoring your marketing strategies to target the right audience effectively.

Reaching out to people who are unaware of your brand or even unaware that they need your product is a significant challenge in sales. This approach, often called cold prospecting, can be time-consuming and less effective, as it involves approaching potential customers without any prior relationship or context.

Maximizing your time by prospecting better is crucial to stack the deck in your favor. This means focusing your efforts on leads that are more likely to convert.

Here are some ways to do this:

- ❖ **Qualifying Leads**: Efficiently qualifying leads helps focus your efforts on potential customers with a higher chance of purchasing. Qualification criteria can include factors like their need for your product, budget, authority to make purchasing decisions, and timeline for purchasing.
- ❖ **Utilizing Referrals and Existing Networks**: Leveraging your existing customer base for referrals can lead to warmer leads. People are more likely to consider a product or service that comes recommended by someone they trust.
- ❖ **Market Research and Targeting**: Understanding your target market profoundly enables you to tailor your sales approach to meet your potential customers' needs and preferences.

Establishing and nurturing relationships, even with those who may not currently be in the market for your product, creates a foundation for future sales. People remember positive interactions and are more likely to turn to you when they are ready to purchase. The sales environment is dynamic, and what works today might not work tomorrow. Continuously learning from your experiences, staying abreast of market trends, and being willing to adapt your strategies are crucial to maintaining a favorable environment for sales.

Marketing

To understand selling, you need to know marketing. Or how to target prospects. To **Get Sales Faster**, you must prospect for the most qualified prospects. Marketing to your prospects will help you find the most qualified leads.

Definition of Marketing: *the process of identifying, anticipating, and satisfying customer needs and wants through the creation, promotion, and distribution of products or services. It involves analyzing consumer behavior, developing strategies to promote and sell products or services, and building customer relationships. Marketing generates customer interest and creates demand for products or*

services, including market research, advertising, public relations, branding, sales, and customer service. Ultimately, marketing aims to increase a business's sales, revenue, and profits.

Use a marketing department or use your skills. Generating a lead or the bare minimum of qualifying and prospect. To know if you can even speak to them. But first, you need to understand who your market is to sell to. You need to market your brand and yourself aside from your marketing department.

How Marketing Works

Let's look at how marketing works. How I tie marketing into sales and how I view the marketing concept is from my point of view as a salesperson. You might have a marketing department, but that doesn't relieve your duties as a marketer. Understanding the flow of marketing helps you sell more. The goal is to use marketing techniques from your position because we are salespeople, not marketers.

Let's say you are selling men's balding cream. Specifically formulated for men and women, it wouldn't benefit it. It may be a bad example, but bear with me. Baldness is more than just a hair loss problem, especially in men. It's an issue intertwined with identity, self-confidence, and societal perceptions. For many men, baldness can lead to insecurity, anxiety, and a feeling of helplessness. Given the deep emotional and psychological connections, selling a product targeted at male baldness is not just about presenting a solution—it's about understanding the nuances of the audience's needs and crafting a compelling narrative around it.

In this situation, you have two types of prospects: your target market (level one) and your secondary market (level 2).

Level One Prospects

Your primary focus is selling to men because of your specific product. This demographic is your target, the bare minimum to qualify.

Level Two Prospects

It is not focused on women, but women will likely know a man who needs the products so that they will buy them for them. You don't want to forget women because they are buyers, too, but they are your secondary market. You need to have a different angle to market and sell to them.

Level two prospects will require a different focus and take away from your primary target audience.

Have you ever noticed that in a retail store, more women are in the men's clothing section than men? Apparently, women like to shop for their men, and they want them to look good. Or men are not very fond of shopping. Either way, this is not a bad idea because women know what they like, and men need to appeal to their target audience. However, in a retail store, they appeal to their primary market of shoppers (women) but have the products for men (secondary market) in their stores. This makes everyone happy.

> **Level One and Level Two prospects can go hand in hand, but you must identify and separate them to develop a plan to appeal to both levels.**

Local Marketing Strategy

The key to improving your reach and winning over new customers lies in understanding why your current customers choose to do business with you and what keeps them satisfied. Leverage this valuable insight to attract fresh customers by speaking to them in a way that resonates with their needs. Use local marketing to understand why your existing customers bought from you and what they are happy about. Use this information to gain new customers.

You need to get known, and you need to think in terms of multiplication. Becoming a recognizable presence in your market is crucial, and you should approach this with a mindset of exponential growth. Where everyone seems to have advice on what doesn't work, valuable information is often about what does work — even though it's rarely shared. That's because many people genuinely don't have a clear answer. They may say things like:

- ❖ "Sending mailers doesn't work." But does it hurt you?
- ❖ "Sending emails doesn't work." But does it hurt you?
- ❖ "Reaching out on LinkedIn doesn't work." But does it hurt you?
- ❖ "Cold calling doesn't work." But does it hurt you?
- ❖ "Cold outreach doesn't work." But does it hurt you?

The bottom line is this: Embrace the power of local marketing. Learn from your current customers, and let that knowledge guide your approach to gaining new customers to stack the deck in your favor. Don't be discouraged by methods that may seem ineffective; small actions will contribute positively to your overall efforts.

Start Where They Are

Imagine you're embarking on a journey with a potential client. Instead of handing them a map and telling them where to go, you ask them about their current location and what interests them. You're essentially meeting them where they are, acknowledging their unique perspective and needs. This is the essence of asking better sales questions and effective positioning. They are starting where your clients are, which means taking the time to understand their current situation, challenges, and aspirations. It's about listening actively and empathetically. Doing so lets you gather valuable information and create a connection based on genuine interest.

Once you've grasped their starting point, you have the foundation to create interest. This involves crafting questions that uncover their pain points and spark curiosity about your solutions. It's not just about addressing their needs; it's about piquing their interest and demonstrating how your product or service can make a meaningful difference in their journey.

This approach transforms the sales process from a one-sided pitch into a dynamic conversation. It's no longer about convincing; it's about collaborating. By starting where they are and creating interest, you build trust and lay the groundwork for a mutually beneficial partnership.

> "It's not about having the right opportunities. It's about handling the opportunities right." - Mark Hunter

Prioritizing your efforts and focusing on the leads most likely to convert is essential. This may mean focusing on leads that have expressed a strong interest in your product or service or have a specific pain point that your offering can address.

Better Prospecting

Cold calling isn't the only critical aspect of generating meetings, but it can be challenging to get prospects interested. Before making a cold call, thoroughly research the prospect and their company. Understand their needs, pain points, and industry challenges. Tailor your pitch to show that you understand their situation and can provide value.

Be yourself during the call and maintain a respectful tone when reaching out cold. Avoid using aggressive sales tactics and aim to establish a genuine connection with the prospect. Building rapport is crucial to increasing the likelihood of a meeting. During the conversation, actively listen to the prospect's responses and address

any objections they raise. Demonstrate that you value their input and are willing to address their concerns.

Follow-up meetings are the key to your success. Instead of attempting to make a sale during the initial call, focus on setting up a follow-up meeting or call. Express your interest in learning more about their needs and how you can help. Don't focus too much on an elevator speech; if you don't know what you will say, you need some script until you know what to say. Continuously practice your cold calling script and be open to making improvements. Record yourself and listen to the calls to identify areas for enhancement. Refine your pitch based on what works best.

You won't often get a response, but you can't just give up either. If you are still waiting for a response after the initial call or email, keep going. Follow up persistently, but not aggressively, using different communication channels like email or LinkedIn.

Prospecting Routine

If you're diligent and create a routine, you might be able to track the time spent prospecting. This is why routine is so necessary. If you have the discipline to block out the time for prospecting, you should know precisely how long you've been prospecting and how much time you put in. It's blocked out on your schedule because it is in your routine.

Once you have a routine set, you'll know exactly how much time you spend on each activity. You need the time to prospect, but what counts is not only the amount of time spent on the activity but also how effective you are with your time within the activity.

You can measure the activities that you are doing to get better results, but that is just the activity. Working on your scripts, learning how and what to say, and increasing your skill level are improvements that cannot be measured. Your improved results are what can be calculated based on progress. You will not get better results if you do not build trust, listen empathically, and learn how you can solve problems. You can measure how many calls you make to how many deals you close over time, but it doesn't mean your KPI will automatically improve because you are measuring it. The two are not necessarily correlated. The behind-the-scenes skills, the skills that can't be measured, are what needs to get better. The fact that you did more does not automatically mean that you are getting better.

For instance, you can make one hundred calls and set no meetings. You can also make one thousand calls and set no meetings. You didn't improve, but your KPI says you will arrange more appointments if you make more calls. Nothing improved except the number of calls you made. The intention of doing more to get more will be assumed because you are working on getting better while doing more. If the KPI you are measuring is not improving, it will be assumed that you are not working on the skills that cannot be measured, which should result in more sales.

"Success in sales is not just about the numbers; it's about the nuances. It's not the quantity of your calls, but the quality of your connections."
- Anonymous

This is why soft skills and competencies are so critical. Increased activity alone will not necessarily translate into improved measurable results. Measuring how much more trust you gain with a prospect would be impossible. It's impossible to estimate how much listening you are doing or measure how much you care about the prospect's problems and what you are willing to do to help them solve their problems.

It's impossible to measure how much motivation you have to get up every day and work toward achieving your goals. Results can be measured and improved. Actions can be measured and improved. The skills needed from step to result cannot be met.

Use the Prospecting Data

How do you qualify more prospects? Use the numbers game to track data to improve your skills, refine your scripts, and refine your strategy. Use the data to track how many "NOs" until you get a "YES." Numbers don't lie. We need to look at your calls-to-close (CTC) ratio. Your CTC tracks how many calls you'll make and how many closed sales you get.

If I were your manager, I would ask you the following:

- ❖ How many calls did you make this week?
- ❖ How many meetings did you set?
- ❖ How many sales did you get?

"Sales" can be viewed as whatever the sale needs to be. A sale could be a closed deal, or a sale could be setting a meeting. Whatever the goal is for that call is the

sale. Upon review, it is determined that you've made ten calls and had one deal. Congratulations! Now, make ten more calls! The idea is to look at the prospecting data and know that you must make ten calls to get a sale. If you made only nine calls, make one more. Over time, the data will create accurate numbers and improve your skills.

Again, If I were your manager, we would go over some of this information during a review to help you improve your CTC.

- ❖ Are you contacting the right people?
- ❖ What was discussed?
- ❖ What is being said to try to close your sales?
- ❖ When will they be ready to buy?
- ❖ How can you improve your scripts to improve your data?
- ❖ Someone will accept your offer, and when they do, why? What went well, and how can you repeat the process?

With coaching and feedback, it is easier to improve your skills. Without tracking the correct data, it's impossible to improve. We can then dive even deeper. How many calls did you make to a particular prospect before you got a sale?

"What gets measured gets improved." – Peter Drucker

Prospecting aims to get answers and get to know your market and prospects. The goal is to get a meeting for the opportunity to cover the first two steps in the sales process: Prospect and Qualify. You can't close the business if you don't move your prospects through the sales process.

Prospecting KPI Data Points

- ❖ Number of calls made to appointments set.
- ❖ Number of appointments closed over appointments set.
- ❖ Number of prospects you have over several contacts made.
- ❖ Number of contacts made to appointments set.

There's also another interesting KPI that we can examine: The number of hours spent prospecting. It's interesting because if you're reluctant to make calls and put the time into prospecting, how likely would you be to track the time you're doing it? I don't know if this is an effective KPI to track because I believe you won't follow the time you put into it if you don't make the prospecting calls. So, I think trying to do that is a waste of time.

Welcome to 2024: Use Technology to Streamline the Process.

Various tools and technologies are available that can help streamline the sales process and make it more efficient. For example, customer relationship management (CRM) software can help you track your interactions with potential customers while marketing automation software can help you nurture leads and keep them engaged over time.

In today's digital age, the right technology can make or break a company's sales efforts. Leveraging technology makes the process more efficient and creates more personalized customer experiences, leading to improved conversion rates.

Technology Streamline

1. **Customer Relationship Management (CRM) Software**
   Example: Salesforce, HubSpot CRM, Zoho CRM

   Why It's Beneficial:

   - ❖ **Organization:** CRM software helps organize information about leads, prospects, and customers in one place. This ensures the sales team doesn't waste time searching for contact details or past interactions.
   - ❖ **Lead Prioritization:** By tracking interactions and collecting data, CRMs can help salespeople prioritize which leads to focus on based on engagement or other criteria.
   - ❖ **Team Collaboration:** CRMs often come with features that allow multiple team members to collaborate on a lead, ensuring everyone is on the same page.

2. **Marketing Automation Software**
   Example: Marketo, Mailchimp, Pardot

   Why It's Beneficial:

   - ❖ **Personalization:** These tools can segment your audience based on behaviors, preferences, or past interactions, allowing for more personalized communication.
   - ❖ **Timely Engagement:** Set up automatic emails or notifications based on triggers, ensuring that potential customers receive timely and relevant content.

- **Performance Analytics:** With built-in analytics, sales and marketing teams can gauge the effectiveness of their campaigns, adjusting strategies in real time.

3. **AI-Powered Platforms**
   Example: ChatGPT by OpenAI

   Why It's Beneficial:

   - **24/7 Customer Interaction:** AI chatbots like ChatGPT can engage customers anytime, providing information, answering queries, or even setting up appointments without human intervention.
   - **Cost-Effective:** Once set up, ChatGPT can handle many customer interactions without the need for continuous staffing, leading to cost savings.
   - **Data Collection:** AI platforms can gather data from customer interactions, providing insights into common queries, pain points, or areas of interest, which can be valuable for refining sales strategies.
   - **Consistent Responses:** An AI chatbot ensures consistent responses, reducing the chances of misinformation or varied answers to the same query.

4. **Communication Platforms**
   Examples: Zoom, Slack, Microsoft Teams

   Why It's Beneficial:

   - **Seamless Collaboration:** These platforms enable smooth communication among sales teams, ensuring quick decision-making and problem-solving.
   - **Remote Selling:** With tools like Zoom, sales pitches and demos can be conducted remotely, catering to global audiences and saving on travel time and costs.

Embracing technology in sales is no longer a luxury but a necessity. Technology plays an integral role in modern sales strategies, from keeping track of customer interactions with CRM software to engaging leads using marketing automation tools and leveraging the power of AI through platforms like ChatGPT. By using these tools effectively, sales professionals can enhance efficiency, foster better customer relationships, and ultimately drive more sales.

Create Curiosity

Get Sales Faster

You need to know how to pique interest and to get a meeting with someone; they need to be interested in meeting with you. Before contacting a prospect, thoroughly research their company, industry, and pain points. Understand their specific needs and challenges so that you can tailor your conversation to address their concerns.

At the beginning of the call, be honest and transparent about your intentions. Clearly state your purpose for reaching out, and let the prospect know that you value their time and are focused on delivering value to them. To gain interest, avoid using a one-size-fits-all sales pitch. Tailor, you're messaging each prospect's unique situation, demonstrating that you have taken the time to understand their business and how you can provide value.

During the initial contact, emphasize your product or service's solutions and benefits. Explain how your offering can help prospects overcome challenges or achieve their goals. Offer something of value upfront, such as free resources, tools, or guides related to the prospect's interests. This showcases your willingness to help without expecting anything in return immediately.

Cold prospecting, reaching out to potential clients with whom you have no prior connection, is a challenging but essential aspect of sales. The primary obstacle in cold prospecting is quickly generating interest or curiosity in someone who might not have an immediate need or awareness of your product or service. To overcome this, it's crucial to start where the prospect is - understanding their potential needs, interests, and pain points. Having a solid idea of what might appeal to them allows you to tailor your approach in a way that resonates.

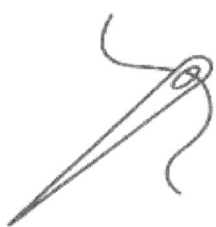
Creating a reason for the prospect to engage with you is the linchpin of successful cold prospecting. This reason should be compelling and relevant to their context. It might be a unique value proposition of your product, a pressing industry trend, or a solution to a common problem they might face. The reason serves as a bridge, connecting the prospect's current state to the potential value you can offer.

Many things go through their mind when reaching out cold over the phone or approaching a stranger to start a conversation. Remember, your prospect must analyze:

- How much time will this call take?

- Do I have time for this?
- Do I feel like doing this?
- How they feel about the decision.
- How they feel about you.
- Answer in their mind if they say yes; what does that mean for them?
- Will it be a waste of my time?
- Is there really a benefit?

If you have something worthwhile that can help people, it's in their best interest to hear more. It's just that people don't have the expertise or knowledge of what you do.

Human psychology suggests that people are more likely to engage when they see an apparent reason or benefit. It's important to articulate this reason succinctly and effectively in your initial communication. This could be in the form of a question that sparks their interest, a statement that highlights a unique benefit, or an intriguing fact that relates to their business.

In addition to providing a reason, personalizing your approach can significantly increase the chances of capturing a prospect's interest. This means doing your homework to understand the prospect's business, industry, and potential challenges. Personalized communication shows that you are not just reaching out randomly but have a specific and potentially valuable proposition for them.

Cold prospecting requires a strategic approach to create interest in a short amount of time. By clearly understanding the prospect's potential needs and interests and presenting a compelling, personalized reason for them to listen, you can effectively pique their curiosity and open the door for further engagement. This approach transforms the cold prospecting challenge into an opportunity to connect with new clients by directly addressing their needs and offering relevant solutions.

The Power of NO

No is the first reaction anyone has. It's a defense mechanism, so embrace it, don't fight it. You want to ask questions to get your customers to open up. Continuously asking your client yes questions may not be the best strategy. Have you heard of "Yes Momentum?" It's a sales philosophy where you build up your clients to say yes through the close.

They can feel correct if backed in a corner and forced to say yes to your customer. The idea is that customers need to figure out the information for themselves and

discover that your solution is the best option for them. If they don't find this, you're not the best option. They will start to say yes to anything to get you to leave.

In his book, *"Never Split the Difference,"* Chris Voss introduces three types of "yeses": the Confirmation Yes, which seeks agreement; the Commitment Yes, which aims for commitment after a series of confirmations; and the Counterfeit Yes, a false agreement. Voss advises against using yes-oriented questions, as they can diminish rapport. Momentum selling, a strategy of obtaining consecutive yeses leading to commitment, is criticized for undermining autonomy and damaging relationships. The "Counterfeit Yes" concept is explained, where a yes masks a no. The preferred approach is suggested to be using No-Oriented Questions, highlighting the power of "no" in negotiations. **People are more comfortable saying no. The trick is to get them to say no while telling you yes. Ask, "Are you opposed" to questions? Change the way you ask questions.**

No Orientated Questions and Position Statements

Q: Are you opposed to seeing how …. (This is really a yes question.)
A: No, I am not opposed …

OK, here's more.

YES to NO Questions Examples

INSTEAD OF...	CONSIDER THIS...
Do you want to see more?	Are you against seeing more?
Can I show you?	Would it be wild to consider?
Do you have time for it?	Is tomorrow a bad time?

Calls-To-Close Ratio

Calls-to-close ratios are a critical metric for measuring a sales team's success. A calls-to-close ratio represents the number of calls a salesperson takes to close a deal. This ratio is essential because it indicates how effectively a salesperson converts leads into customers. It also helps companies to identify areas where their sales team can improve and optimize their sales processes. For clarity, Calls are defined as interactions by Phone, email, or in person.

Why is Tracking Call-to-Close Ratios Important?

Tracking calls-to-close ratios is essential for several reasons. Firstly, it provides insight into how successful a sales team is in converting leads into customers. By

monitoring this metric, sales managers can identify which team members are performing well and those who may require additional training.

Secondly, tracking call-to-close ratios helps companies identify inefficiency in their sales processes. If a particular product or service consistently has a lower call-to-close ratio, it may be time to re-evaluate the sales process for that specific product or service. By identifying these areas, sales managers can optimize their sales processes, reduce costs, and improve overall efficiency.

Finally, calls-to-close ratios help sales teams set realistic goals and targets. By tracking and analyzing past performance, sales teams can create accurate forecasts for future sales and develop strategies to meet their targets.

Strategies for Improving Call-to-Close Ratios.

There are several strategies sales teams can implement to improve their call-to-close ratios. Below are some of the most effective:

- ❖ **Qualify Leads Properly:** Sales teams must spend time qualifying leads properly before making a sales call. Qualifying leads involves gathering information about prospects to determine whether they fit the offered product or service. This will ensure that the sales team only spends time on prospects likely to convert into customers, resulting in a higher call-to-close ratio.

- ❖ **Build Relationships:** Sales is about building relationships. Salespeople must establish a rapport with prospects to gain their trust and confidence. By building relationships, salespeople can create a connection with prospects and better understand their needs, allowing them to tailor their sales pitch accordingly.

- ❖ **Improve Product Knowledge:** Salespeople must have in-depth knowledge of the products and services they are selling. This knowledge allows them to answer questions and provide solutions to problems, which builds trust with prospects and improves the chances of closing the deal.

Get Sales Faster

Tracking calls-to-close ratios is an essential metric for measuring a sales team's success. By monitoring this metric, sales managers can identify areas for improvement and optimize their sales processes. Implementing strategies such as properly qualifying leads, building relationships, addressing objections, and improving product knowledge can help sales teams improve their call-to-close ratios, increasing sales and revenue. In a competitive sales environment, offering value from the first interaction is essential. Sales reps often seek advice on how to provide insights, solutions, or resources during the initial contact that will make prospects more receptive to setting up a meeting.

Chapter 5
Be a Market Maker.

To become a Market Maker, you must navigate a path paved with the foundations of effective sales communication. It's not just about presenting products or services; it's about crafting compelling stories, fostering trust, and cultivating connections that yield profound results. This journey involves understanding the nuances of persuasion, mastering the craft of negotiation, harnessing the power of emotional intelligence, and getting the market on your side.

Entering a new market and creating awareness is not for the weak! You need to be cut out for sales to enjoy the fruits of a market maker. With no understanding of the inappropriate need for your product, prospects will be reluctant to buy. This is why the consultative approach is the most effective. When in the complex sales process, you must create interest in what you have.

> *"The best way to predict the future is to create it." - Peter Drucker*

People in new markets or even old ones don't know their problems. Or at least not the ones that you solve. It would be best to create interest in what you have to offer, and starting conversations with people is the first and most important thing you can do to make this happen. People don't know you, so why should they listen to you? Be curious about how you can help them change.

The A Player

There are two types of people in sales. "A Players" and "B Players". You have guessed it; you want to be an "A Player." To be an "A Player," you need to know what an "A Player" is and what it will take to become one.

"A Players" are the top-performing salespeople who consistently excel and achieve outstanding results. They are often considered the best of the best in the sales profession. "A Players" always meet or exceed their sales quotas and targets. They are the top revenue generators in the sales team and consistently deliver exceptional results. They possess a strong work ethic and are highly motivated. "A Players" are dedicated to their work, often putting in extra hours and effort to achieve their goals.

To be an "A Player," you need to produce! They are highly focused on achieving results and are driven by personal and team success. Companies often value and reward A players because they significantly contribute to revenue growth and overall success.

"A Players" have a deep understanding of the products or services they are selling. They are knowledgeable about their offerings and can confidently answer customer questions. This product knowledge will give you the confidence to understand your products and help your prospects. Knowing your product until it becomes automatic will provide you with an edge during sales conversations because you will be able to help. However, A Players know their customers better. They go the extra mile for their customers. They excel in interpersonal and communication skills. "A Players" effectively builds rapport with customers, understands their needs, and persuades them to make purchasing decisions. "A Players" prioritize the needs and satisfaction of their customers. They aim to build long-term relationships and understand that satisfied customers can lead to repeat business and referrals.

With more opportunities come more problems. "A Players" are, of course, skilled problem solvers, and this is where product knowledge is critical. Since A players know their customers well, they can identify pain points and provide solutions tailored to each customer's unique situation. "A Players" have thick skin. Not everyone will say yes. They are used to rejection because they ask tough questions and can handle rejection and setbacks without losing motivation. They bounce back quickly from challenges and continue to pursue opportunities.

You have to stay sharp to be an "A player." They are committed to self-improvement and ongoing learning. "A Players" stay up-to-date with industry trends and sales techniques to maintain their edge. They can adapt to changing market conditions

Get Sales Faster

and customer needs. "A Players" are flexible and can adjust their sales strategies to stay competitive.

The bottom line is that "A Players" will outwork their competition!

REE-ROO… REE-ROO…

Life as a "B player" can be challenging. "B Players" aren't bad at selling. They just aren't quite there yet. They often lose to the competition or provide the wrong product solution to their customers that doesn't quite fit their needs, so they are passed on. Or they don't quite know their customers well enough, so they don't sell their needs; they sell their products or what they want to sell them, not what the customers want to be sold.

The term "ambulance chaser" comes to mind. "Ambulance Chaser" describes a salesperson who aggressively and unethically pursues potential customers in vulnerable or distressing situations. The term is derived from the practice of lawyers known for chasing ambulances to solicit clients who have recently been injured in accidents.

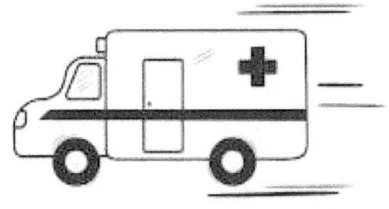

In sales, an ambulance chaser might target individuals who have experienced a recent personal tragedy, financial crisis, or other challenging circumstances to sell them products or services. This could involve exploiting people's emotional vulnerabilities or pressuring them into making quick decisions, often for the salesperson's financial gain rather than the customer's best interest.

Ambulance chasing is generally frowned upon and considered unethical in sales and marketing because it prioritizes sales over potential customers' well-being and informed choices. Ethical sales practices focus on providing value, building trust, and ensuring customers make informed decisions that benefit them.

Being an ambulance chaser can also result in a lack of creativity. I've been in a situation where I've created a good relationship, and the opportunity got lots of attention. There's a lot of transparency in municipal sales, so this particular deal was published in the paper. Now, here comes the fun part. B players like to take advantage of this situation. You have to anticipate this and ensure your relationship is vital. The B player will find every opportunity to derail your deal and make your

prospect second guess their decision. In turn, I use this situation to my advantage. If you know that there is a possibility that this will happen, make sure to coach your customers on what those B players will say before they have a chance to tell them themselves.

In my experience, they over-promise, underdeliver, and try everything they can to undercut their competition. This is a form of negative selling. Instead of trying to sell what they do differently and better, they negatively sell their competition on what they can't do or how they're not a good fit. Luckily for me, the deal was iron, and the relationship was even more robust, so my customer was skeptical, to say the least. People I never heard of or reaching out to my customer for a shot at the deal that was too far along as it is.

When you get desperate as a salesperson and can't get creative, it's not a good look from the prospect's point of view. You need to do a better job of creating your own opportunities. Generating your own needs and creating curiosity for new prospects instead of trying to pick up where someone else has already started. Don't fear this because it's predictable, and you gain more credibility because you know what you're talking about. Don't be an ambulance chaser.

Consultative Selling

Developing a new market often requires educating potential customers about the problems they might need to know about and the solutions available. A consultative selling approach enables you to provide valuable insights, industry knowledge, and educational content to raise awareness about the challenges and opportunities in the market. Positioning yourself as an expert and sharing relevant information can create awareness and establish thought leadership within the new market.

A consultative selling approach and creating awareness while developing a new market are closely correlated. Consultative selling is centered around understanding the customer's unique needs and challenges. Understanding the potential customers and their pain points is crucial when entering a new market. By engaging in consultative conversations, asking probing questions, and actively listening, you can uncover the specific needs and gaps in the market.

You can identify market opportunities within the new market segment through consultative selling. By deeply understanding the customer's requirements, preferences, and aspirations, you can assess which aspects of your product or service are most relevant and valuable in meeting their needs. This helps create

awareness of your offering in a way that resonates with the target market. This information can be used to serve your other customers.

This approach gets them from point A to point B and begs the question: what is the desired outcome of the sale? Bring the prospect's problems to light to get around the "This is the way we have always done things" objection. Address their current state and show them what could be. Ask, "What if things were different? Painting a picture for a prospect of your vision for their future state is way more effective than showing a prospect a product or pointing out their problems with no fundamental understanding of how to fix them.

Consultative selling involves ongoing customer dialogue to gather feedback and understand how your offering can continuously meet their evolving needs. When entering a new market, this feedback loop becomes even more critical. By maintaining open lines of communication, you can gather market insights, identify emerging trends, and adapt your strategies to stay ahead in the market. This iterative process helps create awareness and refine your approach as you develop the new market.

Timing: The Tickle List

The key is more touches and understanding what your customers need and why they need it.

Timing means "at the right time" and includes the right environment. There is a time and place for everything. The time may also be correct, but not necessarily the perfect environment. There may be a time when the environment is right, and you must use emotional intelligence, cues, and some situational awareness to identify when you're in the right setting at the right time for the sale.

Creating the right environment will give you a better opportunity to execute the sales process, and a better time will be present. You don't need to wait to be in the perfect place to manifest itself; you can also create the right environment. For example, you can invite your prospect to a social or sports event. Or if people aren't social, you can get them one-on-one for lunch or in their office. The goal is to position them where they are at ease and receptive to business discussions. Tailoring the optimal environment hinges on understanding their personality type.

You may not even talk about business and get to know your prospects in their environment. Building a relationship instead of "always be closing."

When is the Right Time?

Get Sales Faster

Let's discuss timing during the sales process and how your offering aligns with the customer's needs. Timing is crucial in the sales process, particularly concerning the customer's needs. Aligning the timing of your offering with the customer's need is essential for a successful sale.

Timing is only sometimes predictable, and circumstances can change. Be prepared to adjust your approach and be flexible in meeting the customer's evolving needs. This might involve offering customized solutions or adapting your product to align with the customer's timing preferences.

To understand timing, you must build strong relationships and maintain regular communication with potential customers so you can stay on their radar when the timing becomes right. This can involve keeping in touch, sharing relevant industry insights, or providing valuable content.

As discussed in "Get Good at Sales," not everyone is ready to buy but will when they are ready. Different customers have distinct buying cycles, which can vary based on factors such as industry, budget cycles, and internal decision-making processes. Like buying a car, for instance. The average time a person owns a car in America is 14 years. People will buy a car when they are ready. Gathering information about the customer's buying cycle is essential to identify the optimal timing for approaching them.

Once you have identified the timing of the customer's need, it's crucial to communicate the value of your offering clearly and concisely. Highlight how your product or service can address their pain points and provide immediate benefits, demonstrating that the timing suits their situation. By understanding the customer's pain points and challenges, you can determine when they will most likely require your product or service.

For instance, if your offering addresses a specific seasonal issue or a problem that arises during a certain period, timing your approach accordingly can be highly effective. Keeping track of industry trends and market dynamics can help you anticipate when customers need your offering more urgently. For example, if a new regulation is about to be implemented, you can proactively position your product as a solution to address compliance requirements. This is better done before the law is put into place, so your products are at the top of mind with your prospects.

The timing might need to be corrected because your customers may view their situation as acceptable, and change is unnecessary. Changing people's minds and persuading them to do something different is precisely why **To Get Better at Sales** was written and why there is a sales profession. You can create urgency, manifest the timing, and create a business opportunity that makes customers realize they have an issue. This isn't easy money, and it takes time, effort, and organization to create markets and get people to change.

Timing and creating business opportunities require a deep understanding of the customer's pain points, effective communication, and building trust. By aligning your offering with their needs, providing valuable insights, and demonstrating the urgency of the issue, you can increase the chances of the customer realizing they have a problem and that your solution is right for them.

Create Events

Create an event to get many people to see your message at once. It would be best if you were known to do this; otherwise, no one will come. Or have an offer that is intriguing for people to listen to. I want to start you in the right direction with this. The idea is to get people that warm and in front of you. Not all events will fit into your industry, budget, or customer base. That's not the point; the point is to create an event, even if it's only for one person.

Expand your reach and gain more leads. Organize webinars, live Q&A sessions, or virtual events on social media platforms to share valuable insights and connect with potential prospects in real time. Webinars can be an effective way to generate interest and inquiries.

> Let's take a look at some strategies to create events.

1. Build Credibility and a Draw: To attract a significant audience to your event, it's essential to establish credibility and create a draw. Here's how:

- ❖ **Expertise**: Position yourself or your organization as an expert. Share valuable insights, case studies, or success stories that showcase your knowledge and credibility.
- ❖ **Intriguing Offer**: Develop a compelling offer or message that piques the interest of your target audience. Make sure your event provides a solution to a problem or fulfills a need for attendees.

2. Special Keynote Speaker: A special keynote speaker can be a game-changer for your event. Here are some ideas:

- **Industry Influencer**: Invite a well-known influencer or thought leader in your industry as a keynote speaker. Their presence can attract a significant audience interested in their expertise.
- **Celebrity Guest**: If relevant to your industry or topic, consider bringing in a celebrity guest who aligns with your event's theme. Their star power can generate buzz and attendance.
- **Customer Success Story**: Sometimes, your best spokesperson is a satisfied customer. Consider featuring a customer who has achieved remarkable success using your product or service as your keynote speaker.

3. Offer Value Beyond Sales: To make your event appealing, focus on delivering value beyond sales pitches:

- **Educational Content**: Plan sessions or workshops that provide attendees with actionable knowledge and skills they can apply in their lives or businesses.
- **Networking Opportunities**: Create networking sessions or panels where attendees can connect with peers, experts, and potential collaborators. Networking can be a significant draw.
- **Entertainment**: Incorporate entertainment elements such as live music, performances, or interactive experiences to make your event memorable and enjoyable.

4. Free Event with Tons of Value: Offering a free event can be an excellent way to attract a broad audience. However, ensure that attendees perceive high value.

- **Content Quality:** Focus on providing high-quality content, workshops, or resources that attendees can't easily find elsewhere.
- **Giveaways and Prizes:** Consider offering prizes, giveaways, or exclusive discounts for attendees to enhance the perceived value of the event.
- **Access to Experts**: Arrange Q&A sessions or one-on-one consultations with experts in your field. Attendees will appreciate the opportunity to get personalized advice.
- **Post-Event Resources**: Provide attendees with post-event resources, such as presentation materials, recorded sessions, or additional content reinforcing the event's value.

Remember that successful events require careful planning, promotion, and execution. Consider effectively leveraging digital marketing, social media, email marketing, and partnerships to reach your target audience. Combining these strategies and elements allows you to create events that draw in a large crowd, leave a lasting impression, and effectively drive your message home.

Cognitive Biases

Charlie Munger, the business partner of Warren Buffett, has identified and discussed various cognitive biases that can affect decision-making. He developed his list of cognitive biases based on his extensive experience as a successful investor and businessman. He drew upon his observations and research from various disciplines, including psychology, economics, and statistics.

Munger has stated that he was inspired by psychologist Robert Cialdini, who identified six principles of persuasion in his book "Influence: The Psychology of Persuasion." Munger believed these principles were vital to understanding people's decisions and could be applied to investing and other business areas.

Munger also drew on the work of other prominent thinkers, such as Amos Tversky and Daniel Kahneman, who pioneered the field of behavioral economics and identified various cognitive biases that affect decision-making.

Munger's list of cognitive biases has been refined over the years through his ongoing study of psychology, business, and human behavior. He has emphasized the importance of understanding these biases for making rational decisions and avoiding common pitfalls. Munger's list has become widely recognized in business and investing as a valuable tool for improving decision-making and achieving better outcomes.

Charlie Munger arrived at his understanding of these cognitive biases through his observations and insights from various fields of study. Munger is known for his multidisciplinary approach to decision-making, which draws on principles from psychology, economics, and behavioral science, and his extensive experience in investing and business.

Here's some context on how he concluded that these biases exist:

- ❖ **Personal Experience:** Munger has had a long and successful career as an investor and business leader. Over the years, he has observed how cognitive biases can affect decision-making in finance and business. He

has seen firsthand how investors and businesspeople often make irrational choices based on these biases.
- ❖ **Collaboration with Warren Buffett:** Munger is the longtime business partner of Warren Buffett, and their partnership has involved extensive discussions and analyses of investment decisions. Together, they have witnessed the impact of cognitive biases on investment choices and market behavior.
- ❖ **Study of Behavioral Economics:** Munger has shown a deep interest in behavioral economics, which is the study of how psychological and emotional factors influence economic and financial decisions. Behavioral economics researchers like Daniel Kahneman and Amos Tversky have provided significant insights into cognitive biases, and Munger has incorporated these insights into his thinking.
- ❖ **Interdisciplinary Learning:** Munger is known for his wide-ranging reading habits and ability to draw wisdom from various disciplines. He has studied psychology, history, biology, mathematics, and other fields, which has enriched his understanding of human behavior and cognitive biases.
- ❖ **Mental Models:** Munger often talks about the importance of developing a latticework of mental models, which are mental frameworks that help individuals understand and navigate the world. These mental models are rooted in psychology and behavioral science principles and serve as tools to recognize and counteract cognitive biases.
- ❖ **Learning from Mistakes:** Munger emphasizes the importance of learning from one's mistakes and the mistakes of others. He has seen how cognitive biases can lead to costly judgment errors, which has reinforced his belief in the existence of these biases.

Munger's insights into cognitive biases are theoretical and based on a lifetime of practical experience and continuous learning. His ability to identify and address these biases has been a critical factor in his success as an investor and a decision-maker. His writings and speeches, including those found in "Poor Charlie's Almanack," are valuable resources for anyone looking to improve their decision-making abilities and avoid common cognitive pitfalls.

These cognitive biases are discussed in the context of decision-making, behavioral economics, and psychology, and Munger uses them to illustrate the challenges people face in making rational and effective decisions. Munger's insights into these biases are valuable for anyone seeking to improve their decision-making skills and avoid common pitfalls in thinking.

Charlie Munger, the vice chairman of Berkshire Hathaway and Warren Buffett's business partner, discusses various cognitive biases in his book "Poor Charlie's Almanack."

Charlie Munger's 25 Cognitive Biases

1. Confirmation Bias: The tendency to seek and interpret information in a way that confirms one's existing beliefs or opinions.

    Example: Someone who believes a diet is effective may only seek out and remember information supporting their belief while ignoring or dismissing any conflicting evidence.

2. Incentive-Caused Bias: The bias arises when financial or other incentives lead to a distorted view of reality or unethical behavior.

    Example: A salesperson who receives a commission for selling a specific product may be tempted to exaggerate its benefits or downplay its drawbacks to make more money.

3. Bias from Over-Influence by Social Proof: The tendency to conform to the behavior or opinions of a group, even if they are not rational or correct.

    Example: People standing in a long line for a restaurant may assume the food is excellent because of the crowd, even if they have no other information about the quality

4. Lollapalooza Tendency: A combination of multiple cognitive biases that can lead to extreme and unexpected outcomes.

    Example: The 2008 financial crisis was a lollapalooza of various biases (e.g., overconfidence, herd behavior, incentives) contributing to a massive market crash.

5. Stress-Influence Tendency: The bias occurs when stress or emotional factors cloud judgment and decision-making.

    Example: Making impulsive investment decisions during a stock market crash due to the stress and fear of losing money.

6. Availability Bias: Giving more weight to readily available or easily recalled information, often leading to incorrect conclusions.

   For example, I believe that shark attacks are more common than deaths due to falling coconuts because media coverage of shark attacks is more sensational.

7. Anchoring Bias: The tendency to rely heavily on the first piece of information encountered (the "anchor") when making decisions.

   Example: A seller initially lists a used car for a very high price, which becomes the anchor for potential buyers, making any subsequent lower price seem like a good deal.

8. Mis-wanting tendency: The tendency to misjudge what will make us happy in the future, leading to poor choices.

   Example: Buying a luxury sports car thinking it will make you happy, only to discover that it doesn't bring the expected fulfillment

9. Reciprocation Tendency: The inclination to reciprocate favors or gifts, even when not in one's best interest.

   Example: Feeling obligated to purchase a product from a salesperson who gave you a free sample, even if you don't need it.

10. Deprival Superreaction Syndrome: The tendency to overvalue things taken away from us or become scarce.

    Example: Panic buying and hoarding products during a crisis (e.g., toilet paper shortages during the COVID-19 pandemic).

11. Social-Proof Tendency: Similar to bias from over-influence by social proof, it's the tendency to follow the actions of others in uncertain situations.

    Example: Investing in a trendy but risky startup because everyone else seems to be doing it.

12. Contrast-Misreaction Tendency: Overreacting to differences or changes rather than evaluating them in absolute terms.

Example: Overreacting to a slight salary increase because it's higher than the previous one, even if it's still below market rates.

13. Envy/Jealousy Tendency: Feeling resentment or desire for what others have, leading to irrational behavior.

    Example: Buying an expensive car just because a neighbor got one, even if it strains your finances.

14. Bias from the Non-Mathematical Nature of the Human Brain: The human brain often struggles with complex mathematical concepts and tends to simplify problems, sometimes leading to errors.

    Example: Underestimating the probability of complex events, such as winning the lottery.

15. Bias from Simple, Pain-Avoiding Psychological Denial: Avoiding unpleasant truths or information that may cause emotional discomfort.

    Example: Ignoring signs of a severe health issue because facing it would be emotionally difficult.

16. Bias from Excessive Self-Regard Tendency: Overestimating one's abilities or importance.

    Example: Overestimating your ability to time the stock market and making risky investments.

17. Liking/Loving Tendency: Favoring people, products, or ideas we like or love, sometimes to the detriment of rational judgment.

    Example: Hiring a friend or family member for a job even if they are not the most qualified candidate.

18. Disliking/Hating Tendency: The opposite of liking/loving tendency, where negative emotions lead to biased judgments.

    Example: Rejecting a good idea simply because it was proposed by someone you dislike.

19. Doubt-Avoidance Tendency: Avoid uncertain situations or decisions, even when necessary.

    Example: Staying in a job you hate because you fear the uncertainty of finding a new one.

20. Inconsistency-Avoidance Tendency: The desire for consistency in one's beliefs and actions, even when not rational.

    Example: Continuing to smoke cigarettes despite knowing the health risks because quitting would be admitting a past mistake.

21. Curiosity Tendency: The natural inclination to seek new information or experiences.

    Example: Spending hours researching a topic online, even when it's irrelevant to your immediate needs.

22. Kantian Fairness Tendency: A sense of fairness or justice influencing behavior.

    Example: Feeling compelled to divide a dessert equally among friends, even if one person ate significantly less.

23. Use-It-or-Lose-It Tendency: The belief that something will be lost or wasted if unused.

    Example: Hoarding possessions because you fear that if you get rid of them, you might need them in the future.

24. Survivorship Bias: Focusing on the successes or survivors and ignoring the failures or non-survivors when analyzing data.

    Example: Assuming that successful entrepreneurs must have followed a specific path, ignoring countless failed entrepreneurs who took the same route.

25. Prospect Theory and Loss Aversion: A theory that describes how people value potential gains and losses, often giving more weight to avoiding losses than achieving gains.

Example: Feeling the pain of losing $100 more intensely than the pleasure of gaining $100, leading to risk-averse behavior in investments.

These examples illustrate how cognitive biases can influence decisions and behavior in various aspects of life.

Focus on 5: Create the Lollapalooza Effect

Create a setting with a combination of the five biases. The more cognitive biases that you use, the better during your customer interactions. I have listed five common biases that are simple to use without much effort. Plus, they are the easiest to learn within your target market and how to use them.

- **The Reciprocity Bias:** This bias suggests that people are more likely to reciprocate when someone has done something for them. In direct selling, this means that if a salesperson offers something of value to a potential customer, such as a free sample or a discount, the customer may feel obligated to reciprocate by purchasing.
- **The Anchoring Bias:** This bias occurs when people rely too heavily on the first information they receive when deciding. In direct selling, a salesperson can use this bias to their advantage by presenting a high-priced item first, which may anchor the customer's perception of the product's value. Then, when the salesperson presents a lower-priced thing, the customer may perceive it as a bargain and be more likely to make a purchase.
- **The Social Proof Bias:** This bias suggests that people are more likely to follow the actions of others when making decisions. In direct selling, a salesperson can leverage this bias by showing potential customer testimonials from satisfied customers or pointing out the product's popularity.
- **The Authority Bias:** This bias suggests that people are more likely to trust and follow the opinions and recommendations of perceived authoritative or knowledgeable people. In direct selling, a salesperson can leverage this bias by highlighting their expertise, experience, or credentials or pointing out that experts or industry leaders have recommended the product or service. This can increase the customer's trust in the salesperson and the product, making them more likely to purchase.
- **The Availability Bias:** Availability bias is a cognitive bias that occurs when people make judgments and decisions based on the ease with which they can recall or remember specific information or examples. In other words, individuals tend to overestimate the importance or likelihood of events or news that readily come to mind, often because those events are more recent, emotionally charged, vivid, or widely publicized.

> Let's examine these Biases in action. Here is an example value proposition from SaaS Solution X (a fictitious company).

Lollapalooza

We believe in providing exceptional value right from the start. With SaaS Solution X, we offer a free trial period or a generous onboarding package, allowing you to experience the full potential of our software. Our goal is to become the benchmark for excellence in the industry. Join the ranks of other industry-leading companies that have already embraced SaaS Solution X. Our software has a proven track record of success, and we highlight the experiences and achievements of these organizations.

By leveraging the power of social proof, we demonstrate that you're making a wise decision by aligning yourself with our trusted and respected user base. Trust in our expertise and experience. SaaS Solution X is developed by industry experts and thought leaders who deeply understand your business needs. We provide thought leadership content, whitepapers, and speaking engagements to establish ourselves as the leading authority in our field. By choosing us, you can have confidence in our ability to guide you toward unparalleled success. *We offer a unique offering for new customers to try our product for a free trial.* Act now to seize the limited opportunity with SaaS Solution X.

The Breakdown

The use of multiple biases at once will solidify your chances at persuasion. This strategy plays into the principle of dealing with multiple personalities, and you want to relate to all people. You might miss the opportunity or pique interest if you only try to sell your product with one bias. Here is an example value proposition from SaaS Solution X. Let's break down the strategy in simple terms and dissect the value proposition from SaaS Solution X and how they use cognitive biases to persuade their customers to decide in their favor.

> ➢ *Reciprocity Bias: Gain an Unfair Advantage*

With SaaS Solution X, we offer you unparalleled value from the start. We're invested in your success and want to allow you to experience our software's transformative impact firsthand.

Studies show that when someone receives a favor, they feel compelled to reciprocate. By providing a free trial period or a generous onboarding package, we

trigger the reciprocity bias, ensuring that your decision-makers feel obliged to reciprocate with a long-term commitment to our solution.

➢ *Social Proof Bias: Join the Winning Team*

Join the ranks of other industry-leading companies that have already embraced SaaS Solution X.

Humans tend to conform to social norms, seeking validation from others before making decisions. SaaS Solution X harnesses the power of social proof to position ourselves as the trusted choice in the market. We create a compelling narrative demonstrating our software's widespread adoption and positive impact by highlighting the success stories of industry-leading companies already embracing our solution. Join the ranks of these prestigious organizations and leverage the crowd's wisdom to solidify your decision to choose SaaS Solution X.

➢ *Availability Bias: Seize the Limited Opportunity*

We offer a unique offering for new customers to try our product for a free trial. Act now to seize the limited opportunity with SaaS Solution X.

The availability bias suggests that people assign greater value to things perceived as rare or limited in availability. With SaaS Solution X, we create a sense of urgency by emphasizing the limited availability of our exclusive features, pricing packages, or personalized customer support. By instilling a fear of missing out, we drive you to take action and secure our solution before it's too late. Don't let this golden opportunity slip away – act now and experience the competitive advantage that SaaS Solution X provides.

➢ *Authority Bias: Trust the Experts*

Trust in our expertise and experience. SaaS Solution X is developed by industry experts and thought leaders who deeply understand your business needs. We provide thought leadership content, whitepapers, and speaking engagements to establish ourselves as the leading authority in our field. By choosing us, you can have confidence in our ability to guide you toward unparalleled success.

By leveraging the authority bias, we position ourselves as the leading authority in our field. Through thought leadership content, whitepapers, and speaking engagements, we establish ourselves as the go-to source for insights and solutions. Trust our expertise, and let us guide you toward unparalleled success with SaaS Solution X.

> *Anchoring Bias: Setting the Standard of Excellence*

Our goal is to become the benchmark for excellence in the industry.

The anchoring bias suggests that people rely heavily on the first piece of information they receive when making decisions. By positioning SaaS Solution X as the industry standard, we anchor your perception of what high-quality software should be.

The Bottom Line

The value proposition is designed to create a strong case for choosing SaaS Solution X over alternative options. Ultimately, the value proposition intends to accomplish business transformation, customer adoption, customer success, trust-building, and long-term relationships. It positions SaaS Solution X as the preferred choice for businesses seeking to unlock their full potential with the help of cognitive biases. By highlighting the unique benefits and leveraging cognitive biases, the statement aims to increase customer adoption of the software and establish it as the go-to solution in the market.

Through social proof, authority bias, and a track record of success, it aims to build trust and instill confidence in its customers. By positioning themselves as the industry standard and showcasing the achievements of other successful companies, they strive to create a sense of trust in their expertise and the value they deliver. SaaS Solution X is dedicated to ensuring your success. By providing a free trial period or a generous onboarding package, they allow customers to experience the transformative impact of their software firsthand.

The goal is to enable customers to achieve exceptional results and drive their business forward. By delivering outstanding value, addressing customer needs, and helping companies achieve success, SaaS Solution X aims to establish long-term relationships with its customers. They want to be a trusted partner in their customers' business journey, continually providing innovative solutions and supporting their growth.

Important Disclaimer

Remember that these biases should be used ethically and respectfully, focusing on genuinely addressing the needs and values of your buyers. Tailor your messaging and marketing tactics to resonate with their preferences while maintaining authenticity and transparency. Conducting thorough market research and ongoing

customer feedback can provide valuable insights into the biases that resonate most with your target audience.

Using these biases for the forces of good and not evil persuasion is powerful, and when these techniques are used for the wrong reasons, it is bad ju-ju. You must be able to buy what you are selling because you believe in your product. Scams and get-rich-quick schemes last only a short time.

Pro Tips for Building Markets

- **Know your Market:** Know who you serve and how. Define who your ideal customer is and focus on that customer segment within the market you are trying to penetrate. You must know who you are targeting first.
- **Research and Identify Potential Pain Points:** Conduct thorough research to understand your target audience and their pain points. This can happen by asking your current customers why they chose to do business with you. You can also ask customers who passed why they didn't do business with you. This research also involves studying industry trends, analyzing customer feedback, and monitoring market developments. By identifying common challenges or areas of improvement, you can create an opportunity to address their needs.
- **Craft a Compelling Value Proposition:** Develop a compelling value proposition highlighting how your product or service can solve the customer's problem. Focusing on their specific problem, clearly communicate the benefits, unique features, and advantages of your offering. Making it evident why they should pay attention to the issue at hand.
- **Tailor Your Approach to the Customer's Context:** Every customer is unique, so it's crucial to customize your approach based on their specific context. This involves understanding their business, industry, and individual pain points. You can create a sense of urgency and relevance by demonstrating your understanding and aligning your solution with their needs.
- **Build Familiarity:** Position yourself as a trusted advisor by providing thought leadership and educational content through blog posts, articles, whitepapers, webinars, or speaking engagements, offering valuable insights and information about the customer's pain points. By showcasing your expertise, you can demonstrate the importance of addressing the issue and position your offering as the solution.
- **Use Case Studies and Social Proof:** Leverage case studies, testimonials, and social proof to demonstrate how your offering has solved similar

problems for other customers. Highlight success stories and tangible results achieved through your solution. This can help the customer visualize the potential impact on their business and motivate them to address the issue.

- ❖ **Create a Sense of Urgency:** Establish a sense of urgency by highlighting the consequences of not addressing the issue promptly. This could involve discussing missed opportunities, potential risks, or competitive disadvantages. By emphasizing the time-sensitive nature of the problem, you can motivate the customer to take action and realize the need for your solution.
- ❖ **Engage in Consultative Selling:** Take a consultative approach by actively listening to the customer's challenges and goals during the sales process. Ask probing questions to uncover underlying issues they may not know and demonstrate how your product or service can address them effectively. By guiding them through this discovery process, you can help the customer realize the extent of their problem and the value of your solution.

Chapter 6
6 Deadly Sins of Sales.

Success in sales hinges on building strong relationships, understanding customer needs, and effectively closing deals. However, there are several everyday things that sales professionals often need to correct, which derailed their efforts and led to lost opportunities. We will explore the six deadly sins of sales and why they should always be addressed during the sales process. Recognizing and addressing these pitfalls can significantly improve your chances of achieving sales excellence.

SIN #1

Not getting better results:
The "Rocking Chair Effect"

Set yourself up for success. Don't suffer from the "Rocking Chair Effect." This is where there is much action but no movement, like running around your house fifteen times. You went far and might have even broken a sweat but didn't get anywhere. Have clear, concise goals and perform deliberate actions to accomplish your goals.

One of the main pitfalls is focusing on the activity rather than the skill. Let's take running, for example. It's like this: you could run around your house 30 times. All you would do is break a sweat, but you didn't get anywhere. You did a lot of work, and you're tired. Instead, focus on the skill of running and running somewhere to improve your time of getting there.

Avoid the rocking chair effect: *There was a whole lot of movement, but you didn't get anywhere.*

It's all too easy to get caught up in a whirlwind of activity—endless prospecting calls, numerous meetings, and a constant stream of emails. While these activities are undoubtedly essential, they often lead to a common pitfall: mistaking busyness for actual accomplishment.

Recognize your bad habits or the ones that do not accommodate your system for achieving your goals and replace them with good habits that will work in your approach to achieving your goals. Use the goal sheet to set up a system for achievement and operate the system into your routine.

Mistaking activity for accomplishment is a common problem. It's easy for sales professionals to fall into the trap of mistaking activity for achievement. One main reason is that checking things off your to-do list feels good. Feeling accomplished but needing to accomplish something. This doesn't improve your productivity.

Mind Busyness

Can you list all the things you did today but are still in the same spot you started? More activity translates into something other than achievement. You must distinguish activity from accomplishment; Busyness or busy work is just that, BUSY.

Avoid puttering and laziness: *Avoid choosing mundane activities over important ones to feel better about the fact you did something without accomplishing anything.*

Working from home is very difficult because there are tons of things to do around the house that are manageable tasks, like cleaning, vacuuming, laundry, or lawn watering. These things are just not critical. All these things can wait, but they are easy, mindless tasks that make you feel that you've accomplished something but didn't accomplish what mattered at the time.

Effectiveness: *Think about how your actions or behaviors can be practical or if they will be effective.*

Ask yourself –

- ❖ Is this necessary?
- ❖ Will this action help me move to the next level?
- ❖ How effective is this action to accomplish my goal?

Why Sales Professionals Fall Into the Trap

With targets to meet and quotas to reach, it's tempting to believe that being busy equates to being productive. However, this misconception can lead to wasted time, missed opportunities, and subpar results. The following section will explore why sales professionals often make this mistake and provide practical tips for avoiding it.

Also, you may not know better because you feel accomplished after a flurry of activity. Just doing more to get more sounds logical enough, but you need to be doing the right things.

Understanding the Activity-Accomplishment Paradox

The activity-accomplishment paradox is when sales professionals confuse busyness and productivity. It's a natural human tendency to associate being active with making progress, but this can be misleading in sales. Engaging in numerous activities may give the illusion of progress, but more is needed to guarantee meaningful accomplishments or sales success.

> Several factors contribute to sales professionals mistaking activity for accomplishment:

- ❖ **Pressure to demonstrate productivity:** Sales is a results-driven profession, and professionals must establish their productivity constantly. They believe being busy shows dedication and effort, even if the outcomes don't align with their goals.
- ❖ **Focusing on the wrong metrics:** Many sales teams emphasize activity-based metrics, such as the number of calls made, emails sent, or meetings scheduled. While these metrics can be helpful indicators, they don't reflect the quality or impact of those activities on the sales pipeline.
- ❖ **Fear of missing out (FOMO):** Sales professionals want to take advantage of all potential opportunities, so they engage in multiple activities simultaneously. However, spreading oneself too thin can result in a lack of focus and reduced effectiveness.

How to Avoid Mistaking Activity for Accomplishment:

- ❖ **Define clear objectives:** Begin by setting clear and measurable objectives that align with your sales goals. This will help you focus on meaningful outcomes rather than merely staying busy. For instance, your objective

might be to close a specific number of deals or achieve a particular revenue target within a defined timeframe.
- ❖ **Prioritize your activities:** Identify and prioritize the activities directly contributing to achieving your objectives. This ensures that your time and effort are spent on tasks that generate the most impact. Avoid getting caught up in low-value activities that don't align with your goals.
- ❖ **Track and analyze meaningful metrics:** Instead of solely relying on activity-based metrics, track metrics that provide insights into the effectiveness and progress of your sales efforts. Measure metrics like conversion rates, average deal size, customer satisfaction, and sales cycle length. This will give you a clearer picture of your accomplishments and areas for improvement.
- ❖ **Regularly evaluate and adapt.**
  Take the time to evaluate your sales strategies and tactics regularly. Analyze what's working and what's not and make necessary adjustments. Continuous improvement based on data-driven insights will help you maximize your productivity and avoid falling into the activity trap.

Sales professionals must break free from the misconception that activity equals accomplishment. By focusing on meaningful objectives, prioritizing tasks, and tracking relevant metrics, you can avoid wasting time on unproductive activities. Remember, it's the quality, not the quantity, of your efforts that genuinely drives success in sales. Embrace a strategic and goal-oriented approach to sales, and you'll find yourself achieving meaningful accomplishments rather than simply keeping busy.

SIN #2
Not following up:
How do you keep your prospects engaged?

Follow-up is conducted at every aspect of the sales process. You will need to get the prospect through the cycle, not just at the end, as the term suggests. You may need to follow up after every stage of the cycle. Establish scripting that will help you with your message. Follow-up is critical because people don't always commit because they can't, won't, or aren't asked, and you can't assume that just because you had a meeting or a demo or sent a proposal, it's a done deal.

Do not let the prospect hang out. Set the pace for follow-up and control the process. Set firm follow-up dates and subsequent actions to move the approval and buying process along. It is far too easy for prospects to get busy and forget to put your purchase at the top of their list of things to do.

Follow up: *pursue or investigate something further.*

Follow-up isn't "just checking in." It is a strategic process to ensure your prospect is engaged, on track, and moving forward. Sometimes deals stall, and that's okay, but what is the reason?

Send a quick message: I have not heard from you in a while. It seems like we got off track. Are you still interested in moving forward with … [whatever you are working on.]

Stop, Just Checking In!

Stop just checking in with your customers if the process is stalled. Instead, provide value at every stage of the process. Providing value at every stage of the sales process can help keep momentum and build trust with your potential customers. This may involve providing educational content, product demonstrations, or personalized recommendations based on the customer's needs.

The act of "checking in" is familiar to any sales professional, but it's essential to recognize that merely touching base without purpose can feel disingenuous or irritating to customers. The key is to check in and engage in ways that underscore your value proposition, solidify relationships, and help drive the sales process forward.

Understand the Problem: Just "Checking In."

The phrase "just checking in" is, in essence, empty. It lacks a clear purpose, doesn't offer value, and can make customers feel like they're merely a box to be ticked off. Instead, interactions should be about furthering understanding, building relationships, and facilitating the buyer's journey.

Strategies for Meaningful Engagement Educate with Content: Rather than checking in, consider sharing an insightful article, a relevant case study, or even a white paper that relates to the industry or challenges the customer might be facing. This demonstrates that you're attuned to their needs and positions you as a knowledgeable resource.

- ❖ **Offer Product Demonstrations:** If the customer hasn't seen the full breadth of what your product or service can do, now might be the time. Organize a demonstration tailored to their specific needs, ensuring they understand the full potential of what you're offering.
- ❖ **Personalized Recommendations:** Use your knowledge about the customer to provide tailored suggestions. Whether it's a product feature they might not be using or a new offering that fits their profile, such personalized attention can make the customer feel valued.
- ❖ **Seek Feedback:** Instead of checking in, ask for feedback. How are they finding the product or service? Are there areas where they feel there's room for improvement? Such interactions can provide invaluable insights and demonstrate that you value their opinion.

- ❖ **Share Success Stories:** Highlight how other clients have benefited from your product or service, especially in a similar industry or context. Success stories or testimonials can serve as powerful reassurances and might even inspire the customer to explore further aspects of your offer.
- ❖ **Introduce New Features or Upgrades:** If there have been updates or additions to your product or service, this is a great time to introduce them. Make the client aware of the advancements and how they can benefit.

Maintain a Value-First Approach The overarching strategy should be a commitment to providing value at every touchpoint. Whether through educational content, personalized recommendations, or simply listening to the customer's feedback, every interaction should enrich the customer's experience and understanding.

- ❖ **Send Timely Information:** Regularly share industry-specific content, insightful articles, and thought leadership pieces demonstrating your expertise. This positions you as a valuable resource and keeps you on your prospects' radar.
- ❖ **Establish a Social Media Presence:** Participate in relevant conversations, discussions, and groups on social media platforms. Engage thoughtfully with prospects' posts, comments, and questions to establish a relationship and gain visibility.
- ❖ **Take a Personalized Approach**: Use direct messaging to build rapport and understand your prospects' needs. Engage in genuine conversations to identify potential opportunities.

Merely "checking in" no longer cuts it because there is no value. Sales professionals must adopt a proactive, value-driven approach to engage customers meaningfully. By understanding a customer's needs, challenges, and aspirations, salespeople can pivot from mundane check-ins to valuable interactions that build trust and drive the sales process toward a successful conclusion.

Complex Sales and the Customer's Buying Journey

In complex sales, customers often face a more intricate buying process. These purchases involve substantial investments, longer decision-making timelines, and multiple stakeholders. As a sales professional, it's your responsibility to anticipate and navigate the complexities of your customer's buying process. Because they may not even know

> **Here are the critical steps to understanding and guiding your target customer through their buying process:**

1. **Research and Awareness:** In the initial stage, your target customer may need to know their specific needs or the solutions available. Your role is to provide valuable insights and industry knowledge to raise awareness about potential challenges and opportunities.
2. **Needs Assessment**: Help customers identify their pain points and needs. Conduct thorough discussions to uncover their underlying issues, even if they're unaware of them. Act as a consultant, not just a seller.
3. **Information Gathering**: Customers will embark on information-gathering journeys as they recognize their needs. Provide them with the information they need to make informed decisions. This may include sharing case studies, industry reports, or product demonstrations.
4. **Evaluation and Comparison**: During this stage, your target customer will likely evaluate various options, including your competitors. Be prepared to highlight your product or service's unique value and how it addresses their specific pain points.
5. **Building Trust**: Building trust is paramount in complex sales. Nurture your relationship with the customer, provide exceptional customer service, and establish yourself as a reliable and trustworthy partner.
6. **Negotiation and Decision-Making**: Expect negotiations as part of the process. Be patient and open to compromise, always considering the customer's best interests. Help them navigate any internal decision-making hurdles or concerns.
7. **Purchase and Implementation**: Once the customer decides to proceed, assist them through the purchase process, ensuring a smooth transition from decision to implementation. Provide ongoing support to address any concerns or challenges that arise.

Roadblocks to Following Up

Your top priority is not your prospect's top priority. Sales is simple in concept, especially when you follow the process. How effective you are in the process is the proving ground. When you know the rules, you can play the game. Sales will take perseverance and discipline. Follow up to find out what the holdup is. Knowing why your prospect can't commit will teach you how to overcome the objection.

> **Here are some reasons why people don't commit to buying, and follow-up will keep your prospects in your sales cycle.**

- ❖ **Going in Alone.** Nobody decides by themselves. They will need to consult with someone affected by the change or decision.

- **Just Shopping.** Getting a feel for the going rate or where to set a budget. You are keeping your competition honest, getting a second opinion, and justifying their purchase with your competitor.
- **People are busy**: not lazy, just maybe not that important to them at this exact moment. No big deal but remember that they aren't constantly waking up thinking about calling you back. People are also not annoyed continuously when you do follow up. Plus, you will likely lose the sale if you don't follow up. I recommend following up at the risk of "annoying" your prospect rather than waiting for them to call you.
- **People are too nice.** They may feel bad about telling you no or not interested, so the least path of resistance is avoidance. They may initially say they want to get you off the phone or their office. This is where the sales process comes into play. Help your prospect and help yourself by following the process to avoid unnecessary follow-ups. If you have qualified or unqualified during the early stages of the sales process, you will know if you need to follow up or not.

Know When to Cut Bait

You probably don't need to follow up forever or waste your time. Keep tabs on how often you have tried to follow up with a prospect. Note dates, times, and methods of communication. Usually, after five attempts, you will have a good indication of whether a prospect will call you back, but it could take as many as fifteen! This would be up to you and your prospect to determine. Read between the lines and listen to cues indicating whether the prospect is interested in moving forward. Know when to stop wasting energy on a prospect that isn't warranted. Let it cool down for a while and try again. Like fishing for sharks, try different bait if they aren't biting.

SIN #3
Not Recognizing the Small Sale: Trying to close quickly.

Shark fishing in the coastal waters of South Carolina was one of the most fun things I have done. There are no big sharks here, no matter how big or small a shark is. They were blacktip sharks. I have been fishing several times before, even catching some marlin off the coast of Cabo San Lucas.

This shark experience was different and one of the perfect analogies to describe sales and how to make it easier.

To catch these sharks, you need bait. The best way is to use the bait the sharks are already eating. You want to give yourself the best chance to catch the sharks instead of trying to guess.

We started with small bait, prawns, or shrimp for the layperson. At first, I was a little disappointed and frustrated because we were not catching sharks even though we were shark fishing. We were catching whiting, which is a small bait fish. Now, the only reason why I was disappointed is that we weren't catching sharks. That's because I didn't know how to catch sharks. Starting small is a great strategy. Like a foot-in-the-door strategy in sales, I have a bigger goal in mind, but we started small.

Starting with the shrimp, we attracted the whiting, which, in turn, attracted the sharks. This is an excellent analogy because what I've often done in my life is go for the big sharks first. I promise you'll be waiting a long time because this is a failed

strategy. Of course, I'm not saying you can't catch a giant shark immediately, but you have a better chance if you start small.

Another critical component is knowing when to start switching bait after catching the whiting because the whiting stopped biting. This is how we knew that the sharks were in the area. This same observation can apply to a sales strategy. Know when to move to the next level or switch your strategy based on your situation.

Create interest and get the sharks to come to you! The coastal waterways are vast, and you want to go where the sharks are, but you can't see under the water. However, they are full of fish and full of sharks. You need the sharks to come to you. Like in sales, you need to create interest and go where the people are in your market or bring the people to you. Networking!

Small Wins to Big Deals.

Small wins are getting agreement from your prospect to continue. Getting to the next sale, the next meeting will contact you for bigger deals. You have to catch the small fish before landing a shark. You can't expect to get to the top. It takes time for the small sales to add up to the big deals.

> *Great things are done by a series of small things done together.*
> *– Vincent van Gogh*

The same hook, the same equipment, just a different strategy. This can be applied to the sales process. Just sell something minor, like a meeting, and then gain your customers' trust to earn more of their business.

Sell the Next Meeting

Think of the small sale as a test by your customer. This could be an initial meeting. A sale is a sale in that if your prospect agrees to an ask, you made a sale. In the complex sales process, "selling the next meeting" plays a pivotal role, akin to "eating the elephant one bite at a time." This metaphor encapsulates managing overwhelming tasks by breaking them into smaller, more manageable steps.

One bit at a time translates to focusing on securing micro-commitments, such as arranging an initial meeting, which can be seen as small but significant victories in

the broader sales process. These smaller commitments serve a dual purpose. First, they act as trust-building steps, allowing you to demonstrate the value of your product or service and to establish a rapport with the customer. Second, they represent progressive milestones toward the ultimate goal of closing the deal. You cannot just show up and throw up! Your Prospect must have time to digest what you are saying and feel comfortable with the solution presented.

Patience and adaptability are essential in complex sales scenarios, where decisions often involve multiple stakeholders and extended timelines. Each interaction and meeting becomes an opportunity to pitch your product, deepen your understanding of the client's needs, and refine your approach accordingly. The focus, therefore, shifts from trying to close the deal in one go to securing the next interaction.

This approach of selling the next meeting helps maintain momentum and build a cumulative case for your product or service. It's about consistently demonstrating value at every step and nurturing a relationship that transcends the transactional nature of the exchange.

By breaking down the sales process into these more minor interactions, what initially seems like an impossible task becomes a series of achievable steps, ultimately leading to the successful closure of the sale. This method simplifies the complex sales process and lays the groundwork for a sustainable and mutually beneficial business relationship.

SIN #4
Stopping at No:
Persistence Will Pay Off

Sales is a challenging profession, not just because it involves convincing others to believe in a product or service but also because it requires a deep understanding of human behavior. One of the most crucial qualities a salesperson can have is persistence. But it's essential to understand that persistence isn't about being pushy; it's about understanding, listening, and re-approaching with the right strategy.

> *Our greatest weakness lies in giving up. The most certain way to succeed is always to try just one more time*
> -Thomas Edison

No matter how talented a salesperson might be, they'll inevitably face rejection. However, a single "no" doesn't always mean the door is closed forever. Sometimes, it's merely an invitation to try a different approach, offer more information, or wait for a more suitable time. The key is to view rejection not as a defeat but as a temporary setback.

"If the door is locked, try the window."

In many cases, a "no" isn't a rejection of the product or service but rather an indication that the prospect needs more information or has concerns that haven't been addressed. Instead of seeing "no" as a brick wall, consider it a signal to gain more information. Often, a rejection is due to insufficient understanding. A prospect might be unaware of the benefits your product offers or how it can solve a problem they're facing.

In this case, providing additional details or demonstrations can be the key to changing their mind. When faced with objections, it's vital to understand that many are not outright refusals but rather questions in disguise. For example, a comment like "It seems too expensive" might mean, "Can you show me the value this offers over alternatives?" Addressing the underlying question can pivot the conversation in your favor.

Rejection should allow you to regroup and refine your approach.

- ❖ **Listen Actively:** Before you rush to counter an objection, genuinely understand it. Ask open-ended questions and listen to the answers without interrupting. This shows respect and can help you uncover the reasons behind the resistance.
- ❖ **Be Empathetic:** Putting yourself in the prospect's shoes is essential. Understand their needs, challenges, and reservations. You can tailor your pitch to address their unique concerns by doing so.
- ❖ **Stay on Their Radar:** Circumstances can change even if a prospect is not ready to buy now. Regularly check-in (without being intrusive) and provide updates, new information, or industry insights. This keeps the relationship warm and could lead to a future sale.

While persistence is an admirable quality, striking a balance is essential. Being too aggressive can turn off potential clients. The idea is to be resilient, not relentless. Show genuine interest in helping the prospect and be willing to walk away if it's clear the fit isn't right. Your respect for their decision can leave a positive impression and potentially lead to referrals or future opportunities.

Persistence bridges a "no" and a potential "yes." But it's a bridge built on understanding, empathy, and the ability to adapt. Salespeople can unlock the true power of persistence by viewing objections as opportunities and rejections as chances to refine and re-approach.

SIN #5
No Established Next Steps:
"What's Next?"

Clarity and direction are imperative. After a successful pitch or presentation, it's not the end but merely a transition to the next phase: securing commitment and closing the deal. This is the "What's Next?" moment in the sales process.

Be a Tour Guide

I needed a coach or a guide in this case. I have been fishing before but am no expert and have never caught a shark. The guide taught me how to do it. He coached me through the process and settled me down. But most importantly, he taught me patience during the process. You must be a guide, like the guide did for me.

For example, what would be more attractive to you if you wanted to go on a tour and hire a guide to do so? What do you want to see, or what does the guide wish to show you? The simple answer is what you want to see. The guide's job is to understand what you want to see. He can take you on an entire adventure that he knows, like the back of his hand, with the things that interest him, but it would suck because you weren't interested in the things that he wanted to see. He will need to ask you questions so the tour will be a memorable experience for you. On tour, there are one thousand things to see, and he doesn't have time to show you

one thousand different things, so to be a practical tour guide, he needs to ask questions to understand what's the most important to you. That way, you get the most benefit from your tour.

This is like a presentation to sell your stuff. Since you're the guide and you know all the different directions that a display can go in because of all the products or services you can offer up to a prospect, you need to ask the pointed questions to get to the meat of your offering that is most valuable to the prospect.

So, how do you guide your prospect through this critical stage?

- ❖ **Debrief and Guide:** After discussing with your prospect, take a moment to recap. Summarize key points and emphasize what's important to your prospect. Based on the information you've shared, give your expert recommendation. Remember, prospects often look to salespeople as sellers and consultants. Offer your guidance and be assertive about the best next steps. As you rightly mentioned, confidence can significantly impact this phase. When your prospect intends to purchase, lead with a statement showcasing your expertise and assurance, such as, "Fantastic! Here's what we'll do next."
- ❖ **Securing the Commitment:** It's thrilling when a prospect verbally agrees to a sale. However, as you've pointed out, the excitement can quickly fade without a clear path forward. You've "caught the fish," so to speak, but now you need to "get it in the boat." Ensure you are equipped with everything necessary —an order form, contract, or electronic signing platform – to solidify the commitment there and then.
- ❖ **The Importance of Immediate Action:** The time after the prospect agrees is crucial. Leaving without securing the order increases the risk of the deal falling through. Think of it as offering white-glove service: a seamless, high-quality experience without room for doubt or second thoughts. When you have a verbal commitment, turn it into a formal agreement.
- ❖ **Mitigating Potential Roadblocks:** The sales landscape is riddled with stories of deals that seemed confident but somehow slipped away. Prospects are inundated with tasks, priorities, and distractions like everyone else. A week or even a day can shift their focus. An agreement that appeared to be a "no-brainer" can quickly become forgotten or deprioritized.
- ❖ **Always Be Prepared:** One of the best ways to prevent deals from slipping through the cracks is to be prepared. Never attend a meeting without the tools and documents needed to close. Whether it's a physical contract, an electronic signing platform, or a tablet to record orders, ensure you can smoothly transition from discussion to deal completion.

The "What's Next?" phase in the sales process is arguably as crucial, if not more so, than the initial pitch or presentation. It's the bridge between intent and action, between interest and commitment. As sales professionals, mastering this phase means understanding your product and human behavior, building trust, and confidently leading.

SIN #6
Not Closing:
See the deal to the end.

Few actions carry as much weight and significance as 'closing the deal.' Surprisingly, even seasoned professionals sometimes need to improve at this pivotal moment, leading to lost opportunities and diminished credibility. Understanding the importance of closing and why some salespeople miss this step can offer insights into enhancing one's selling skills.

Recognizing the Problem: What is "Not Closing?"

Not closing refers to the failure or reluctance of a sales professional to finalize a sale after initiating and nurturing a relationship with a prospect. It's the equivalent of running a marathon and stopping just shy of the finish line.

> Not closing can manifest in several ways:

- ❖ **Hesitation:** The salesperson delays or avoids asking for the sale due to fear of rejection or a perceived lack of readiness on the prospect's part.
- ❖ **Over-educating:** Continually providing information beyond the prospect's requirements, causing confusion or decision paralysis.
- ❖ **Failure to Address Final Objections:** Not clarifying or resolving the prospect's lingering concerns, leading to missed opportunities.

The Consequences of Not Closing

Closing isn't just a missed sale; it's a series of ripple effects that can harm a salesperson's career and the company's bottom line:

- ❖ **Lost Revenue:** Every sale that doesn't close is revenue left on the table.
- ❖ **Wasted Resources:** Time, effort, and resources spent nurturing a prospect without a resultant sale are essentially wasted.
- ❖ **Diminished Confidence:** Repeated failures to close can undermine a salesperson's confidence, making future sales attempts more challenging.
- ❖ **Eroded Trust:** If a prospect is ready to move forward but senses hesitation from a salesperson, it can erode trust and even raise doubts about the product or service.

Asking for the Sale

Have you ever felt that hesitation when it comes to asking for the sale, or if you've struggled with closing business opportunities? The act of asking for a sale and closing a business can be both exhilarating and intimidating. You've built rapport, understood your customer's needs, and presented your product or service in the best light possible. Now, it's time to seal the deal, but how do you navigate this delicate dance without jeopardizing the relationship you've cultivated?

There's a critical moment when the success of your efforts hinges on your ability to ask for the sale and close the deal. It's the culmination of all your hard work, the product of your persuasive skills, and the point at which a potential customer transforms into a satisfied client. Yet, despite its paramount importance, many sales professionals and entrepreneurs hesitate at this pivotal juncture. Asking for the sale effectively is an essential skill in sales and can significantly impact your success.

Here are some tips to help you ask for the sale persuasively and effectively:

- ❖ **Build Confidence:** This can't be stressed enough. Believe in your product, approach, and, most importantly, yourself. If you're convinced of the value you're offering, it will be easier to instill that belief in your prospects.
- ❖ **Recognize the Right Moment:** Closing should feel like a natural progression of conversation. Be attuned to buying signals such as nodding, agreement, specific product details, pricing, or implementation queries.
- ❖ **Create a sense of urgency:** One effective technique is to create a sense of urgency by emphasizing limited-time offers, discounts, or special promotions. By highlighting that the opportunity is time-sensitive, you encourage the customer to act promptly.

- ❖ **Use assumptive language**: Instead of asking if the customer wants to make a purchase, use assumptive language that implies they are ready to buy. For example, phrases like "When would you like to get started?" or "Shall we proceed with the purchase?" can subtly nudge the customer toward making a decision.
- ❖ **Ask directly and confidently:** When the time is right, ask for the sale directly and confidently. Use clear and concise language, such as "Are you ready to make a purchase?" or "Would you like to proceed with the order?" Maintain a positive and enthusiastic tone while conveying your confidence in the product or service's value.

Closing is an art that encapsulates the essence of the sales process. It's the culmination of understanding a prospect's needs, presenting a solution, and finally transitioning from a conversation to a commitment. By recognizing the pitfalls and arming themselves with strategies and confidence, sales professionals can elevate their ability to close, turning potential misses into confirmed successes.

Trail Closes

Trial Closes are techniques used in sales to gauge a prospect's interest and commitment level throughout the sales process. They are essentially questions or statements that help a salesperson assess how close a prospect is to making a purchase decision. This approach is precious for understanding the prospect's mindset and addressing any potential objections or concerns they might have before the final close.

The purpose of trial closes is twofold. Firstly, they act as checkpoints throughout the sales conversation, allowing the salesperson to assess the customer's readiness to buy and gather feedback on any concerns or objections they may have. Secondly, the format of these trial closes—whether phrased as questions or statements—subtly suggests a commitment, like asking for opinions on the product or proposing hypothetical scenarios about the product's or service's use.

Effective utilization of trial closes involves several vital strategies. Timing is critical; these techniques should be employed at various stages of the sales conversation, not just at the end. This helps in continuously gauging the prospect's mindset. Listening and observing the prospect's responses and body language is vital, as this feedback informs necessary adjustments in the sales approach. Addressing concerns head-on, using feedback from trial closes, is essential, whether providing more information, offering reassurances, or adjusting the offer to suit the prospect's needs better.

Trial closes also facilitate incremental commitment, similar to the concept of selling the next meeting, securing smaller commitments that lead up to the final decision and thus building momentum towards the sale. Customization is another crucial aspect; tailoring a trial close to the specific needs and interests of the prospect demonstrates a thoughtful, personalized approach rather than a one-size-fits-all method. Sometimes, trial closes can create a sense of urgency or FOMO (Fear of Missing Out), nudging the prospect toward a quicker decision. Finally, when used effectively, trial closes can soften the final close, making the concluding step of the sale feel more natural and less pressured, as the prospect has been gradually and skillfully guided toward the decision.

Examples of Trial Closes

- ❖ **Feedback Questions:** "How do you feel about what we've discussed so far?"
- ❖ **Hypothetical Scenarios:** "If you were to choose this solution, how do you see it impacting your daily operations?"
- ❖ **Soft Commitments:** "Would a demonstration next week fit into your schedule?"
- ❖ **Opinion Seekers:** "Do you think this product meets your needs as we've discussed them?"

Chapter 7
6 Essential Skills to Increase Your Sales Success.

Sales is an art form that requires a delicate balance of strategy, persistence, and relationship-building. To truly excel in this competitive field, it is crucial to master the six essential elements that can elevate your sales game to new heights. We will explore these elements and explain why they should always be addressed in the sales process.

ESSENTIAL SKILL #1
Staying Focused:
Chase the Squirrel Later

In every corner of your life, distractions are abundant, and the ability to stay focused can be the difference between mediocrity and extraordinary success. Staying focused is a superpower that every sales professional should harness, as it plays a pivotal role in achieving and exceeding sales targets.

The Price of Distraction

Sales can often feel like a chaotic jungle, teeming with constant stimuli that threaten to pull you away from your goals. Emails flood your inbox, meetings pile up, and the siren call of social media beckons. Amidst this frenzy, maintaining focus is akin to navigating through a dense forest without losing your way. When your focus strays, you risk missing vital opportunities, failing to nurture leads, and losing precious time; distractions come at a high cost.

Here are some of the critical prices you pay when you let your focus slip:

❖ **Missed Opportunities:** Every interaction with a prospect is an opportunity. Whether it's a phone call, email, or face-to-face meeting, each moment presents a chance to move a step closer to a sale. You may

overlook these opportunities when you lose focus, allowing potential clients to slip through your fingers.
- ❖ **Reduced Productivity:** Distractions drain your productivity like a leaky faucet. Constantly switching tasks or being interrupted disrupts your workflow, making it challenging to accomplish essential sales activities efficiently.
- ❖ **Damaged Relationships:** Sales is as much about building relationships as it is about closing deals. Not being fully present during client interactions can lead to misunderstandings, miscommunications, and, ultimately, a breakdown in trust.

Laser Focus

Sales is a highly dynamic, demanding skill and requires unwavering focus. Recognizing the significant benefits that come with laser focus can transform your approach to sales, leading to greater efficiency and success.

> *"The successful warrior is the average man, with laser-like focus."* -Bruce Lee

Focusing intently on a single task enhances your efficiency, allowing you to complete tasks faster and more accurately. This helps you tackle your responsibilities more swiftly and opens up opportunities for engaging in more strategic sales activities. Moreover, active listening, a critical component of effective salesmanship, is greatly improved when you're fully present in your interactions. By staying focused, you're more adept at picking up subtle cues, understanding objections, and identifying client needs, thus demonstrating a deep commitment to solving their problems.

Another key benefit of staying focused is enhanced problem-solving. Sales often involve navigating complex challenges, and with a sharp focus, you're better equipped to methodically address these challenges, finding creative solutions and overcoming objections with greater ease. Furthermore, consistent follow-up, a crucial aspect of sales, is more effectively managed when you maintain focus, ensuring regular nurturing of leads and ongoing communication, ultimately increasing the likelihood of converting prospects into clients.

Staying focused on sales is the key to unlocking your full potential and achieving remarkable results. By recognizing the costs of distraction, embracing the benefits of focus, and implementing effective strategies, you can harness the power of concentration to propel your sales career to new heights.

Example: Imagine you've identified a potential lead, but you need help with administrative tasks, social media, or an unrelated call. The likely lead gets cold, and you miss the opportunity. Always prioritize functions that directly impact your bottom line.

Effective Strategies to Maintain Focus

Implementing specific strategies can be incredibly helpful in harnessing these benefits. Begin by prioritizing daily tasks, focusing first on the most critical tasks that drive sales. Adopting time blocking for specific sales activities like prospecting, client meetings, and follow-ups is also beneficial. During these periods, strive to eliminate distractions and immerse yourself fully in the task at hand.

Mastering the art of focus in sales is about reducing distractions and strategically channeling your attention and efforts where they are most needed. By embracing these strategies and recognizing the value of concentration, you can unlock your full potential in sales and achieve remarkable results. In sales, as in many areas of life, success often hinges on hard work and focused, deliberate actions that drive towards your goals.

- ❖ **Prioritize Tasks:** Start your day by identifying the most critical tasks that will drive sales. Focus on completing these high-priority activities before addressing less crucial matters.
- ❖ **Time Blocking:** Allocate specific time blocks for sales activities, such as prospecting, client meetings, and follow-ups. During these blocks, eliminate distractions and fully immerse yourself in the task.
- ❖ **Mindfulness:** Mindfulness techniques can train your mind to stay present and resist wandering thoughts. Meditation or deep breathing exercises can be valuable tools for maintaining focus.
- ❖ **Eliminate Distractions:** Identify common distractions in your work environment and take steps to minimize them. This might involve silencing email notifications, closing unnecessary tabs on your computer, or setting specific hours for social media use.
- ❖ **Breaks and Rewards:** Set regular breaks in your schedule to recharge. Reward yourself for staying focused by taking short breaks to clear your mind or indulge in a quick treat.

ESSENTIAL SKILL #2
Grit and Grind:
Consistent Action

In life, success rarely comes easy or quick. It takes hard work, perseverance, and dedication to achieve your goals. And while many qualities contribute to success, two of the most important are grit and grind. However, both grit and grind are often associated with working hard to achieve success, but they have slightly different connotations.

"The difference between a successful person and others is not a lack of strength, not a lack of knowledge, but rather a lack in will." - Vince Lombardi

While grit and grind are critical for success, they focus on different aspects of the process. Grit is more about mindset and attitude, while grind is more about the work that needs to be done. Grit helps you stay motivated and committed to your long-term goals, while grind enables you to progress daily.

When you're working towards a goal, you're bound to face obstacles along the way. You may encounter a setback, or you may face criticism or rejection. Whatever the block, it's essential to have grit to persevere and stay committed to your goal. Grit helps you stay focused on the big picture, even when things get tough. Grind enables you to make progress. While grit helps you stay motivated, grind helps you make progress daily. The small, repetitive tasks add up to significant accomplishments over time.

One of the most boring things to do when woodworking is sanding wood. However, it is satisfying to take a rough piece of wood and make it smooth, even with a machine. To do that, though, the process is tedious work.

The sanding must be consistent with long strokes; otherwise, one area would become too low and now level with the rest of the board. You almost need start all over to get the whole board to the same level again. This is "grind" in a nutshell. Not working too hard in any area but being consistent across the board, using grit to your advantage, and not overworking any areas. Ultimately, hard work pays off because you're finished with a nice, smooth piece of wood.

Your grit is like the sandpaper in the example. It is challenging and will make things smooth and cut out rough edges. Be firm to create whatever you are challenged with to make it smooth. Grit refers to the quality of having perseverance and passion for long-term goals and the ability to persist in the face of obstacles and setbacks. It's about having a "never give up" attitude, staying committed to a goal, and being resilient when things get tough.

Grind is the everyday routine to be consistently used with grit to make things smoother. I use grind in the context of mundane, tedious work. Doing whatever it takes to succeed, even sacrificing short-term pleasures or taking on unpleasant tasks. I write of grind as the consistent daily work to make everything happen. Practice, consistency, and routine are all part of the grind. If you are grinding the right way, you shouldn't have to sacrifice the most important things in your life.

In essence, grit is the determination to succeed, while grind is the daily effort required. With grit, it is easier to maintain the focus and resilience needed to stay committed to a long-term goal. And with grind, it's possible to progress towards that goal daily.

Grit and grind form a powerful combination that can help you achieve almost anything. Grit helps you stay motivated and committed to your long-term goals, while grind enables you to progress daily. When you have grit and grind, you can overcome obstacles, stay focused on your goal, and progress daily. So, if you're working towards a goal, remember the importance of grit and grind. Stay committed, work hard, and never give up. With enough grit and grind, you can achieve anything you want.

ESSENTIAL SKILL #3
Follow-Through.
In other words . . . Execution

Where, promises are made, relationships are built, and deals are closed. Execution and follow-through are pillars of success. These actions are not mere formalities but the engines that power the sales process.

Promises Made, Promises Kept

Sales professionals are often seen as dealmakers and persuaders, but their value lies in their ability to deliver on promises. Whether it's a commitment to provide a solution, meet a deadline, or address a client's specific needs, execution is the linchpin that transforms words into actions.

The Cost of Broken Promises

Before we get into the benefits of execution and follow-through, let's examine the consequences of failing to fulfill your promises:

- ❖ **Eroded Trust:** Trust is the currency of success. You erode client trust when you fail to follow through on commitments. They begin to doubt your reliability and question your dedication to their needs.
- ❖ **Lost Opportunities:** Unfulfilled promises lead to missed opportunities. Prospects who feel let down are unlikely to engage with you in the future, and they may share their negative experiences with others, damaging your reputation.

❖ **Stagnated Progress:** A lack of execution can lead to stalled deals and protracted sales cycles. Failure to take action can result in prospects losing interest, causing them to seek solutions elsewhere.

The Benefits of Flawless Execution

The ability to execute your promises flawlessly is not just an asset; it's a necessity. This impeccable execution has numerous advantages that can significantly boost your professional reputation and effectiveness. Consistently delivering on your commitments cultivates a reputation for reliability and dependability. This reliability makes you the go-to choice for clients who value trustworthiness in their sales professionals.

By executing your promises with precision, you not only meet but often exceed client expectations, which is crucial in building and strengthening long-term relationships. Clients who trust your ability to deliver are more likely to remain loyal and become long-standing partners. Moreover, clients who experience your flawless execution will likely become your brand ambassadors. They often enthusiastically recommend your services within their network, leading to an influx of new opportunities through referrals. This word-of-mouth marketing is invaluable and often results in a domino effect of new business.

Another significant benefit is the acceleration of the sales process. When clients witness your efficiency and professionalism, their decision-making process is often expedited, leading to quicker closures and shorter sales cycles. This efficiency not only benefits your clients but also enhances your productivity.

Strategies for Ensuring Consistent Execution

Flawless execution in sales goes beyond mere promise-keeping; it's about creating a legacy of trust and efficiency. You can significantly elevate your sales career by recognizing the value of executing promises precisely and employing strategic planning, clear communication, and continuous improvement. Remember, your actions speak louder than words in sales, and your performance solidifies your standing as a trusted and effective sales professional.

To achieve such impeccable execution, specific strategies are essential:

- ❖ **Clear Communication**: It all starts with deeply understanding your client's needs and expectations. Clear and effective communication sets the foundation for successful execution, minimizing misunderstandings and aligning objectives.
- ❖ **Detailed Planning**: Invest time in thorough planning. Develop a comprehensive roadmap that outlines the steps, timelines, and responsibilities required to fulfill your promises. This plan should serve as a guide to avoid any oversights or delays.
- ❖ **Accountability**: Establish clear accountability for each task. Every team member should know their specific roles and responsibilities, ensuring that all aspects of the commitment are met.
- ❖ **Regular Follow-Up**: Keep the communication channels open throughout the execution phase. Regular updates and responsive support reinforce your dedication to the client's success and help address any concerns promptly.
- ❖ **Continuous Improvement**: Post-delivery, evaluate your performance. Reflect on what worked well and what could be improved. This self-evaluation is crucial for continuous growth and enhancement of your execution skills.

The Importance of Execution in Sales

Evidently, the ability to act rather than just plan distinguishes successful sales professionals from the rest. Execution is the critical bridge between setting goals and achieving them, where the actual value of a sales strategy is realized.

In sales, execution is where the rubber meets the road. Having a well-thought-out plan is one thing, but applying it brings tangible results. The quote by General George S. Patton, "A good plan, violently executed now, is better than a perfect plan tomorrow," perfectly encapsulates this idea. It emphasizes the need for action, even if the plan isn't flawless.

> **Execution in sales involves a series of proactive steps:**

- ❖ **Initiating Action**: The first step in execution is to start. It might seem daunting, especially when faced with an unknown or potential failure. However, as the saying goes, "Get moving before rigor mortis sets in." This means taking that first step, however small, to get the ball rolling.
- ❖ **Overcoming Fear of Imperfection**: Perfectionism can be a significant barrier to execution. The focus should be on progress, not perfection. It's essential to recognize that plans can evolve and adapt. The key is executing and adjusting as needed based on feedback and results.

- ❖ **Following Through with Goals**: Setting goals is only the beginning. The real challenge lies in following through with them. This requires discipline, persistence, and staying focused on the end goal, even when faced with setbacks.
- ❖ **Transforming Hope into Faith**: While hope is a passive expectation, faith is an active belief in one's abilities. In sales, relying on hope means waiting for external circumstances to align in your favor. In contrast, having faith in your execution skills means taking control and making things happen through deliberate actions.
- ❖ **Adapting and Evolving**: A successful execution strategy involves being adaptable. It's about making informed decisions, learning from experiences, and being flexible enough to change course when necessary.
- ❖ **Measuring and Adjusting**: Execution is also about measuring the effectiveness of your actions and adjusting your approach accordingly. This could mean refining your sales pitch, altering your follow-up strategy, or re-evaluating your target market.

Execution in the sales process is about transforming plans and goals into actions and results. It involves starting despite uncertainties, prioritizing progress over perfection, and having faith in your abilities to make things happen. By focusing on execution and adaptability to change, sales professionals can effectively turn their strategies and goals into successful outcomes.

ESSENTIAL SKILL #4
Timing:
Situational Awareness

Mastering the art of timing and situational awareness can mean the difference between sealing the deal and losing a valuable opportunity. Recognizing buying cues, listening to answers to questions, and understanding the importance of these actions in the sales process can significantly enhance your success. You still need to ask for the sales and close even if you can create the timing, as we discussed in Chapter 5.

The Power of Timing

Timing is more than just knowing when to make your pitch; it's about understanding the customer's readiness to buy. Act at the moment when the time is right.

> *"Wherever you are, be there."* – Jim Rohn

Here's why timing is pivotal in sales:

- ❖ **Capitalizing on Urgency:** A sense of urgency can drive decisions in many sales scenarios. Recognizing when a prospect's needs align with the timing of your solution can create a compelling case for them to act swiftly.
- ❖ **Avoiding Missed Opportunities:** Failure to act at the right moment can result in missed opportunities. A prospect needing your product or service but needing to receive the right message at the right time may opt for a competitor's offering instead.

❖ **Strengthening Client Relationships**: Timely follow-ups and communication demonstrate your commitment to meeting the client's needs. This reinforces trust and can lead to long-lasting, mutually beneficial relationships.

The Role of Situational Awareness

Situational awareness in sales involves paying attention to verbal and non-verbal cues, understanding the context of the conversation, and being fully present during interactions. Effectively, Situational awareness allows you to identify potential objections or hesitations from the prospect. This enables you to address concerns promptly and provide information that eases their doubts.

Buyers often drop subtle hints that they're ready to make a purchase. These signals can include questions about pricing, product specifications, or delivery timelines. Attention to these cues allows you to tailor your responses to guide the prospect toward a decision. Understanding the context of the conversation and the prospect's unique situation allows you to personalize your pitch. Tailoring your message to their specific needs and challenges can significantly impact them.

Strategies for Timing and Situational Awareness

Now that we've established the importance of timing and situational awareness in sales, let's explore strategies to hone these skills:

- ❖ **Active Listening:** Pay close attention to what the prospect is saying. Avoid interrupting and focus on understanding their needs, preferences, and pain points.
- ❖ **Asking Probing Questions:** Use open-ended questions to encourage prospects to share more about their situation and requirements. This not only provides valuable insights but also keeps the conversation flowing.
- ❖ **Reading Non-Verbal Cues:** Observe the prospect's body language, tone of voice, and facial expressions. These non-verbal cues can reveal their interest, engagement, or hesitation level.
- ❖ **Tailoring Your Pitch:** Based on the information you gather through active listening and questioning, tailor your pitch to address the prospect's needs and concerns. This demonstrates your attentiveness and commitment to their success.

❖ **Timely Follow-Ups:** After a meeting or conversation, promptly follow up with relevant information or responses to questions. This keeps the momentum going and shows your dedication to serving the prospect.

Timing and situational awareness are your allies in capturing the buying moment. You can elevate your sales game by recognizing the significance of these elements, actively listening, asking probing questions, reading non-verbal cues, and personalizing your approach.

Remember that every prospect is unique; understanding their needs and buying readiness is the key to unlocking sales success. So, stay vigilant, responsive, and attuned to the opportunities that timing and situational awareness can bring your way.

ESSENTIAL SKILL #5
Building Your Network:
Develop Relationships.

The significance of building a robust professional network cannot be overstated. Beyond meeting your immediate sales goals, cultivating a solid network has profound and enduring effects on your long-term career.

Build Relationships

Building strong and lasting relationships with clients and prospects is a cornerstone of success in complex sales. It's not merely about making a sale but cultivating relationships that yield fruitful partnerships and drive repeat business. This section will explore practical strategies and techniques that will help you forge meaningful connections with your clients and prospects.

> *"If you believe a business is built on relationships, make building them your business."*
>
> *– Scott Stratten*

One of the most potent tools for building relationships is understanding. Take the time to thoroughly understand your client's needs, challenges, and goals. The following guide explores practical strategies to forge meaningful connections with your clients and prospects, ensuring long-term success.

Understanding Client Needs

- **Conduct Comprehensive Research**: Before engaging with a client, research their industry, competitors, pain points, and recent developments to demonstrate your commitment and expertise.
- **Effective Communication Skills**: Utilize open-ended questions and active listening during conversations to gather valuable information, showing genuine interest and engagement in their perspectives.
- **Tailor Solutions to Client Needs**: Customize your solutions to meet their unique needs and challenges, showcasing your commitment to their success. This includes needs assessment, highlighting value, and adapting solutions as required.

Maintaining Communication and Building Trust

- **Consistent and Proactive Communication**: Engage in regular check-ins and updates on industry news and solicit feedback to show your dedication to their success and gather insights for improvement.
- **Responsiveness and Reliability**: Ensure timely responses and follow through on commitments. Address issues promptly and proactively to demonstrate your problem-solving skills.

Deepening Relationships Through Personalization and Value

- **Personalization in Communication and Offers**: Remember critical personal and professional details and adjust communication styles to match client preferences.
- **Delivering Continued Value**: Regularly review product/service performance, provide training and education, and offer exclusive deals or loyalty programs.
- **Emotional Connections and Authenticity**: Show empathy, celebrate successes together, and maintain authenticity to build deeper, more enduring relationships.

Advanced Relationship Management Strategies

- **Trusted Advisor Role**: Position yourself as more than a salesperson by offering strategic guidance, aligning solutions with clients' long-term goals, and mitigating risks.
- **Conflict Resolution and Feedback Management**: Develop conflict resolution skills, be open to feedback, and strive for win-win solutions.
- **Partnership and Collaboration**: Foster a culture of partnership through shared goals, joint planning, and co-creation of value.

Leveraging Technology and Measuring Success

- ❖ **Use of CRM and Automation Tools**: Implement CRM systems, utilize automation for routine communications, and leverage analytics for client insights.
- ❖ **Measuring Relationship Health**: Establish KPIs like client satisfaction surveys, Net Promoter Score, and client retention rates to gauge and improve the health of client relationships.

Building and maintaining client relationships in complex sales is an ongoing process. The effort invested in understanding, communicating, personalizing, and nurturing these relationships will yield dividends in the form of loyal clients and successful partnerships.

The Power of a Strong Network

At the core, a strong network is a goldmine for lead generation. You unlock a world of referrals and warm introductions by establishing solid relationships with colleagues, clients, and industry peers. This boosts your immediate prospects and lays the groundwork for future opportunities. In addition, being part of a diverse network keeps you informed about the latest industry trends and market insights, ensuring you stay ahead in the competitive landscape.

Mentorship is another crucial element accessible through your network. Connecting with experienced professionals provides valuable guidance and accelerates your learning curve, facilitating personal and professional growth. Furthermore, your network is a breeding ground for collaboration opportunities, where partnerships and joint ventures can flourish, leading to innovative business endeavors.

Long-Term Advantages of a Robust Network

Over time, the benefits of maintaining a dynamic network manifest in various forms. Continuous interaction with a broad spectrum of professionals fosters ongoing growth, keeping you well-informed and adaptable. Your reputation within your network also strengthens, marked by reliability and integrity, significantly enhancing your professional standing.

A well-maintained network is also a source of resilience during challenging times. It provides practical solutions, advice, and the emotional support essential for

navigating downturns. Moreover, as you engage and contribute to your network, you gradually build a legacy that transcends immediate job roles or sales figures. Your name becomes synonymous with excellence and dependability, influencing far beyond your direct interactions.

Building and nurturing a strong professional network is not a mere career strategy; it's an investment in your future. It's a source of opportunities, a platform for growth, and a support system that can guide you through every stage of your career. The power of networking lies in its ability to open doors, provide insights, and create lasting relationships that foster immediate success and long-term achievements.

Strategies for Building Your Network

Building your network in sales is not just about making connections; it's about investing in your long-term career success. A robust professional network provides many opportunities, insights, and support that can propel you to new heights. So, start today, and remember that your network is not just about who you know but also about who knows you and what you can bring to the table. Building a network takes time, effort, and authenticity.

> **Here are strategies to help you construct and maintain a robust professional network:**

- ❖ **Attend Industry Events:** Conferences, seminars, and trade shows are excellent opportunities to meet like-minded professionals. Make an effort to attend such events regularly.
- ❖ **Online Networking**: Utilize professional networking platforms like LinkedIn. Engage in meaningful conversations, share valuable content, and connect with industry peers.
- ❖ **Join Associations**: Many industries have associations or organizations facilitating networking. Consider becoming a member and actively participating in their activities.
- ❖ **Offer Value**: Be a resource within your network. Share insights, offer help when needed, and make meaningful contributions. The more you give, the more you receive.
- ❖ **Maintain Relationships**: Building your network is a collaborative effort. Nurture relationships over time by staying in touch, expressing gratitude, and offering support when appropriate.

ESSENTIAL SKILL #6
Positioning:
Go where the puck is going to be.

The famous quote, "I skate to where the puck is going to be, not where it has been," is attributed to Wayne Gretzky, one of the greatest hockey players ever. While it's widely associated with him, the exact origins of the quote may not be as clear-cut as they seem.

The quote reflects Gretzky's exceptional ability to anticipate the movements of the puck and his opponents on the ice. It speaks to his incredible hockey IQ and knack for being in the right place at the right time.

Gretzky's playing style was characterized by his exceptional vision and understanding of the game. He didn't rely solely on his speed or physicality but instead used his intelligence to outmaneuver opponents. He could read the game so well that he seemed to know where the puck would be before it even got there.

Anticipate Where the Sale Will Go

Positioning your message strategically is like laying the foundation for a successful conversation. It's about setting the stage for a productive discussion that engages

your prospects and helps them see your value. One crucial aspect is preempting the sale and guiding your customer's thought process and expectations before diving into the pitch.

Imagine you're conversing with a potential customer, and you abruptly launch into your sales pitch without any context or preparation. Their natural response may be to put up their defenses, feeling caught off guard or even tricked. This is where preempting the sale comes into play.

By preempting the sale, you're essentially leading your customers to where you want them to be mentally. You're putting them in a mindset that prepares them for the upcoming discussion, making them more receptive to your message. This ensures that their brain isn't overloaded with information and prevents them from feeling blindsided.

A classic example of effective salesmanship involves selling refrigerators to Eskimos. On the surface, it may seem impossible, given that Eskimos live in frigid conditions, and keeping things warm is the last thing on their minds. However, a skilled salesperson reframes the conversation. Instead of focusing on the refrigerator's ability to keep things cold, they highlight its capacity to maintain warmth. Not all cold foods can be frozen.

In this example, the salesperson doesn't just push the product; they shape the customer's perception. They successfully preempted the sale by guiding the Eskimo customer to understand the refrigerator's value in a different context. This demonstrates the power of strategic positioning and reframing.

So, as a sales professional, remember that your ability to preempt the sale and set the stage can significantly impact your success. It's about leading your prospects toward a clear understanding of the benefits of your product or service, making them more likely to say "yes" when the time comes to seal the deal. By strategically positioning your message, you create a win-win situation where your customers find value, and you achieve your sales goals.

Pre- Framing

Pre-framing or setting the tone for the sale is one of the most important things you can do. Setting the foundation for the deal to be built on getting the customer in the right frame of mind, understanding the process, and having a clear vision of the end Wall helps me set up for sale. You want to tell a story and answer questions before they are asked.

You have to anticipate what the customer might ask and fill in the gap for them. You want to control the narrative because you risk the customer making assumptions. The prospect might not ask the question at all, then make up their answer from their assumption and quietly base their decision on their belief. It's not good because they are making assumptions. You want to ask clarification questions to check for understanding of important aspects of your offer.

Take Away Things That Can Be Objections.

You are selling your house. An intelligent realtor told me less is more when selling your home—the screen door example. You may think a broken screen door is better than a no-screen door because it is still there and provides more features than you have in the house. This is not true.

You may do more harm than good if you try to sell your house with a broken screen door. The buyers will demand that the screen door be fixed before they buy it, or it may deter people from buying it because of the extra work. You are better off removing the door like it wasn't even there, and then it's not an issue. No one will say," You need to put a screen door on here for me to buy it." They will say, "You must fix that door before I buy it."

Taking away uncertainty or not having one little thing in your deal that can create more work for you to fix is another strategy to avoid objections. You can also address the issues right away and work with the customer. "Yeah, I know the door is broken. Would you like me to fix it for you or take it off so you don't have to worry about it?" The door, in this example, is in every deal. There seems always to be one little thing that someone gets hung up on.

Preempting Objections to the Sale

Addressing objections before they happen seems like common sense. But is it? Now that you read that, it seems like common sense, but you need a strategy. That strategy starts with what types of questions you are getting or what type of pushback (objections) you are getting. After you have a solid idea of these, tailor your value position, elevator speech, and presentation to address as many questions and objections as possible before your prospect can consider it. Once they have the information, it is a non-event. If you know your stuff, they will have less to object to.

I can't possibly know all of the objections that you will face. It is appropriate that most complaints are the same, just worded differently. The most commonly asked

sales questions vary depending on the industry and context. You must have answers and defend these objections before the customer can ask them.

Here are 20 customer-originated sales-related questions that are frequently asked:

1. What does your product/service do?
2. How can your product/service benefit me or my business?
3. What sets your product/service apart from the competition?
4. How much does your product/service cost?
5. Are there any discounts or promotions available?
6. Can you provide references or customer testimonials?
7. What is your company's experience in this industry?
8. How long does it take to implement your product/service?
9. What kind of support or training do you offer?
10. Are there any additional fees or hidden costs?
11. Can I try your product/service before making a purchase?
12. What is the warranty or guarantee policy?
13. How does the payment process work?
14. Do you offer customization options?
15. Can you provide case studies or success stories?
16. What is the return/refund policy?
17. Are there any limitations or restrictions I should be aware of?
18. How scalable is your product/service?
19. Can you explain the implementation process?
20. What kind of ongoing support can I expect after the sale?

These questions are just examples and may not apply to every sales situation. This exercise is to know what you must address during sales conversations. It's essential to tailor your questions based on your specific product or service and the needs of your potential customers.

Conclusion:
Connecting the Dots

We examined several strategies and tips for implementing those strategies to **Get Sales Faster**. Consistent action must be included in putting all these actions and strategies together. You must regularly keep up with the actions to **Get Sales Faster**.

Routine and consistent compound actions are the bedrock of success in sales and many other aspects of life. The importance of these principles cannot be overstated. Sales is not just about closing deals; it's about building lasting relationships with clients and customers. Fostering trust and credibility is easier with a routine that involves regular outreach, follow-ups, and consistent efforts to understand customer needs.

Moreover, compound actions, small and consistent efforts that accumulate over time, are critical to sustained success. Like compound interest grows your financial wealth, compound actions build your reputation, knowledge, and skills.

Whether in sales, building relationships, or continuously learning, embracing routine and consistent compound actions is the surest path to achieving your goals and ensuring long-term success.

Routine

Who has an assistant? Not me and not you! You need a routine or, in other words, a system.

Out with the bad habits and in with the new. Settling into a routine is like setting up a system for your life. Setting a routine will keep you honest and allow you to compound your efforts. Establishing a routine will help you achieve your targets, improve your marks, and ultimately achieve your goals once you know precisely what you will do and when you will get things done.

Block out time to get your calls in, answer emails, exercise, eat, or whatever. It's your routine, not mine.

Routine: *a sequence of actions regularly followed; a fixed program.*

Consistency: *the achievement of a level of performance that does not vary significantly in quality over time.*

A consistent, deliberate routine will lead to habits. Habits will confirm the way and lead to action. Being consistent will allow your efforts to compound. Making calls for an hour a day is just that one hour. What else can you schedule?

Time can and will get away from you. Do you feel like you are pulled in twenty different directions? Well, get on a routine. Do you know what you are going to do before you do it? Have ten emails to answer? Wait to answer them until your scheduled time! They can wait an hour or two. Nobody needs an answer immediately. If it's an emergency, they will call you.

With a routine, it is easier to gain momentum. If you are rolling on prospecting calls and get disrupted, it is tough to get the motivation to get started again; you lose momentum. Without the routine, you could fall into a trap of chaos because you are all over the place. With too much to do or a lack of focus, you can easily accomplish nothing.

Entropy: *lack of order or predictability; gradual decline into disorder.*

If you are a traveling salesman, what does Tuesday look like? Then what will a Tuesday look like on the road? Your morning routine doesn't have to change. Your evening routine doesn't have to change. You can have it the same way in different environments. Being consistent will help you be productive. Being productive will help you with purpose. Having a purpose will help you achieve your goals.

Small actions every day will compound like interest in money investments. It can compound both negatively and positively. How can you do the little things over a long period to bring value to our customers, your profession, and yourself? Once you have the habits down, please pay it forward. Help your family grow. Help your friends succeed. Help your customers grow. Help your connections grow.

> *"If you do small actions daily and increase them, they will become big results."*

Compounding is the process whereby interest is credited to an existing principal amount and to interest already paid, i.e., interest on interest.

How?

Help yourself before you help anyone else. Here, we will focus on you. This is one key area where you will need the most help. Take care of yourself so you can always care for others.

Implementing compound actions daily requires

- Accountability
- Systems
- Routine

Why?

You can only get something from something; getting something worthwhile takes time. You need to maintain sight and focus on what you want to get. You will need to set goals to obtain anything you want. Even if you think winning the lottery is a good strategy because it's free money, that's fine, but you still need to work.

You will still need to make money first to buy the lottery tickets, then pick numbers, and finally go to the store to buy lottery tickets. It would be best if you still work to achieve your goal. These tasks are relatively easy, but your chances of winning are next to impossible, so little work equals little payout.

> *"How much you DO will determine how much you GET."*

1. Problem-Solving and Market Understanding:

- ❖ **Problem-Solving Approach**: Position your message to set up meetings based on the problems you solve for your prospects. Highlight how your product or service addresses their pain points.
- ❖ **Knowing Your Market**: Understand your market and level-one prospects thoroughly. Dive deep into their needs and challenges to craft a compelling sales pitch.

2. Communication and Objection Handling:

- ❖ **Effective Communication**: Communication is critical in sales. The right angle to sell a product or service often involves deeply understanding your prospect's needs.
- ❖ **Pre-framing**: Setting the tone for the sale through storytelling and anticipating questions helps prospects understand the process and builds their confidence.
- ❖ **Compelled Preempting the Sale**: Address objections before they arise. Tailor your presentation to tackle common objections and questions. Be proactive in providing information.

3. Common Sales Questions:

- ❖ **Understanding Objections**: Recognize the objections you commonly encounter and strategize to address them in your sales pitch.
- ❖ **Adapting to Different Questions**: While objections can vary, they often share similar wording. Be prepared to handle objections effectively by knowing your product or service inside out.

4. Tailoring Sales Conversations:

- **Customization**: Customize your sales pitch based on the specific needs of your potential customers. This ensures that your message resonates with them.

By incorporating these concepts into your sales approach, you'll be better equipped to understand your market, communicate effectively, preempt objections, and tailor your sales conversations for success.

The Sales Professional's Role

By understanding your customers' needs and pain points, you can tailor your sales approach to their unique requirements. As a sales professional, you play a multifaceted role in the sales process:

- ❖ **Educator**: Act as an educator by providing valuable insights and knowledge. Your expertise can guide the customer's decision-making process.
- ❖ **Consultant**: Adopt a consultative approach, focusing on the customer's needs rather than just selling a product. Ask probing questions to uncover their specific pain points.
- ❖ **Problem Solver**: Be a problem solver by offering tailored solutions that address the customer's unique challenges. Showcase the benefits of your product or service in solving their problems.
- ❖ **Guide**: Act as a guide throughout their buying journey, helping them navigate complexities and make informed decisions.

Your role as a sales professional extends beyond selling; you're a guide, consultant, and trusted advisor. By offering exceptional support and guidance, you can increase your chances of making a sale and build lasting relationships that benefit you and your customers in the long term.

To **Get Sales Faster** in a complex sales process with long sales cycles can be challenging, but there are several strategies that you can use to help speed up the process.

- ❖ **Identify key decision-makers and build relationships**: In a complex sales process, multiple decision-makers may be involved in the buying process. It's essential to identify who these decision-makers are and build relationships with them. This will help you understand their needs and concerns and tailor your approach accordingly.
- ❖ **Focus on the most promising leads:** In a long sales cycle, it's essential to prioritize your efforts and focus on the leads that are most likely to convert. This may mean focusing on leads that have expressed a strong interest in your product or service or have a specific pain point that your offering can address.
- ❖ **Provide value at every stage of the process:** Providing value at every stage of the sales process can help keep momentum and build trust with your potential customers. This may involve providing educational content, product demonstrations, or personalized recommendations based on the customer's needs.

- ❖ **Stay in touch and follow up regularly:** In a long sales cycle, staying in touch with your potential customers and following up periodically is essential. This will help keep your offering top of mind and ensure that you can address any questions or concerns that may arise.
- ❖ **Use technology to streamline the process:** Various tools and technologies are available to help streamline the sales process and make it more efficient. For example, customer relationship management (CRM) software can help you track your interactions with potential customers while marketing automation software can help you nurture leads and keep them engaged over time.

ABOUT THE AUTHOR

With two decades of thriving in sales and sales management, I'm here to share the experience and wisdom I found to work for me. I'm not your typical success story. There was no silver spoon, no secret shortcuts, and nothing handed to me on a silver platter. I've rolled up my sleeves, put in the hard work, and crafted my path to success. And I believe you can do it too.

My mission is simple: to guide you on your journey to success, just as I've learned to navigate on my own. I've ventured into uncharted territories, breathed life into new markets, and defied expectations in various industries with massive success. I've learned along the way that you don't have to go it alone. Success is often a team effort, and I'm here to be part of your team. Whether you're forging a new career path or seeking to level up your existing one, I've got insights, strategies, and a treasure trove of knowledge to share.

My passion for understanding people led me to earn a psychology degree from Northern Illinois University. The intricacies of the human mind have always fascinated me, and this understanding has been a cornerstone of my success. I have always wanted to help people. After all, sales are fundamentally about connecting with people and helping to meet their needs.

In sales, it's easy to feel lost or unsure of your direction, especially when you need more support. My goal is to empower you and as many others as possible. I've worn many hats, walked in different shoes, and understand the challenges you might face. Whether you have a supportive mentor or are trying to forge your path solo, my program is designed to help you connect the dots and chart your course to success.

Join me on this journey of growth, self-discovery, and achieving your full potential. It's never too late to take the first step toward building your desired future.

Together, we'll unlock the doors to your success story.

ABOUT PERSONAL DEVELOPMENT WORKSHOP

Welcome to Personal Development Workshop – your ultimate destination for unlocking health, wealth, and happiness secrets. We're all fueled by something that propels us forward; for me, it's the transformative power of personal development.

Personal Development Workshop isn't just a platform; it's a movement. I've discovered immense fulfillment in my personal growth journey, and my goal is to help you find the same. Through the blog, I share insights into my educational journey, personal improvement, and topics that pique my curiosity, all to help you achieve your best self.

In every blog post, you'll discover practical wisdom about everyday behaviors and habits that lead to a fulfilled life and a legacy that endures. We're here to connect the dots on your journey to success and self-discovery. Join us at Personal Development Workshop, and let's embark on this transformative journey together.

Thank you for purchasing this book. I hope you find value in the ideas and can implement them into your sales process immediately!

Let us know how it is going for you.

Visit our website at www.personaldevelopmentworkshop.com and start your quest for a better, happier you!

www.ingramcontent.com/pod-product-compliance
Lightning Source LLC
Chambersburg PA
CBHW072143170526
45158CB00004BA/1486

Netduino 2 en Español

CARLOS RODRIGUEZ NAVARRO

Copyright © 2014 Carlos Rodríguez Navarro

All rights reserved.

ISBN: 1505435226
ISBN-13: 978-1505435221

AGRADECIMIENTOS

Este libro quiero dedicárselo a mi familia pues sin su ayuda y apoyo nunca lo hubiera podido escribir…

Contenido

1 INTRODUCCIÓN 9

    1-Introducción a la plataforma Netduino 9

    2-Netduino 2 12

    3-Preparación del entorno en Microsoft ® Windows 7 16

    4-Preparación entorno en Microsoft ® Windows 8 17

2 Ejemplo "Hola mundo" 25

    1-Introducción 25

    2-Creación de un nuevo proyecto 29

3 LED parpadeante 33

    1-Introducción al uso de salidas binarias 33

    2-Ejemplo de aplicación que emplea el led interno 34

4 Entradas y Salidas digitales 37

    1-Introducción a las entradas binarias 37

    2-Ejemplo de aplicación que usa el pulsador interno 39

5 Modulación por ancho de pulso 43

    1-Introducción al PWM 43

    2-Ejemplo de aplicación con PWM 44

6 Sensor de ruido 49

    1-Procesamiento de señales analógicas 49

    2-Ejemplo de aplicación que usa una señal analógica 51

7 Sensores de posición 55

    1-Aplicación de los sensores de posición 55

    2-Ejemplo de aplicación que usa un sensor de posición 56

8 Sensor de Temperatura 61

1-Introducción 61

2-Ejemplo de aplicación 64

3- Sensores DHT11 y DHT22 67

9 Sensores de luz 69

1-Introducción a los LDR 69

2-Ejemplo de aplicación con un LDR 71

10 Sensor de movimiento 75

1-Introducción a los PIR 75

2-Ejemplo de aplicación con PIR 78

11 Sensor de consumo eléctrico 83

1-Monitorización del consumo eléctrico 83

1.1-Medidas de seguridad 83

1.2-Optimización del consumo eléctrico 83

1.3-Medida del consumo eléctrico 85

1.4-El estándar DIN 43864 87

2-Conexionado de un vatímetro digital 87

3-Ejemplo de aplicación 88

4-Uso de Interrupciones con Netduino 93

12 Manejo de un display LCD 99

1-Introducción al estándar HD44780 99

2-Ejemplo de aplicación con un LCD 101

13 Servidor avanzado 111

1-Servidor NeonMika.Webserver 111

2-Configuración de red de Netduino 2 Plus 113

3-Métodos soportados en NeonMika Web server 113

GESTIÓN DE SEÑALES DIGITALES 114

GESTIÓN DE SEÑALES ANALÓGICAS 115

UTILIDADES 117

14 MIT App Inventor 121

1 Instalación de MIT App Inventor 121

2-Pasos para crear una aplicación 128

Paso1: Diseño interfaz 128

Paso 2: Adición de la lógica 129

15 Aplicación móvil que interactúa con N2+ 141

1-Introducción 141

2- Diseño interfaz gráfico 144

3- Bloques lógicos de la aplicación 148

4- Aspecto final 160

ANEXO1: Actualizar Netduino 165

Restablecer configuración de red 172

ANEXO 2: ERRORES 175

1-Error 0x8D131700 al compilar 175

2-Cómo corregir el error 10060 175

3-Qué hacer si no se puede compilar su código 177

ANEXO 3 181

ANEXO 4 183

ANEXO FINAL 187

1 INTRODUCCIÓN

1-Introducción a la plataforma Netduino

Netduino es una potente plataforma abierta basada en Microsoft® .NET Micro Framework con una clara visión open source, que además es compatible pin a pin con la famosa plataforma Arduino. Además, al igual que con Arduino, también están disponibles todos los archivos de diseño y el código fuente.

Gracias al uso de Microsoft®. NET Micro Framework, se combina la facilidad de un lenguaje de programación de alto nivel (C #) con un moderno y potente entorno de desarrollo, como es Microsoft® Visual Studio 2010, disponible gratuitamente a través de la siguiente url: http://www.visualstudio.com en su versión Express. Este entorno permite la depuración de programación basada en eventos, multi-threading, ejecución línea a línea, inserción de puntos de interrupción y mucho más, como veremos en este libro.

Respecto al hardware actual de Netduino, está basado en un potente micro-controlador de 32 bits integrado Cortex® a 168MHz con 384kb para almacenamiento y al menos de 100Kb de RAM, lo cual lo hace muy superior a otras plataformas como Arduino (un Atmel® AVR® que funciona a 8 ó 16MHz) o Mega Arduino (que funciona a 16Mhz).

	Arduino Uno	Mega Arduino	Netduino 2
Velocidad del procesador	8 o 16 MHz	16MHz	168MHz
Código Almacenamiento	32kB	256KB	384KB
RAM	2KB	8KB	100 KB +
Entorno de Programación	Arduino (sintaxis similar a C)	Arduino (sintaxis similar a C)	C #, arquitectura net, mono nativo en desarrollo
I / O	14 digitales, 6 analógicas / digitales	54 digitales, 16 analógicas	14 digitales, 6 analógicas / digitales
Corriente CC por Pin I / O	40mA, en total máximo 200mA	40mA, en total máximo 200mA	25mA, en total máximo 125mA
Tensión de I / O	5V	5V	3.3V, 5V tolerante
Precio	~ $ 30	~ $ 59	~ $ 60

En cuanto al hardware, éste permite una fácil interconexión con interruptores, sensores, ledes, dispositivos serie, etc. gracias a que Netduino combina en su interior un GPIO con SPI, I2C, 4 UART (1 RTS / CTS), 6 PWM y 6 canales de ADC.

Además, gracias a los "escudos" (placas de la misma medida que se enchufan directamente en los conectores de la parte superior), es posible añadir más accesorios y periféricos debido a la gran variedad de éstos para Arduino, que son también compatibles con Netduino. Estos escudos ofrecen funcionalidades extra, como por ejemplo, la ubicación GPS, el control de servos, conectividad Ethernet, salidas de alta potencia, ampliación de memoria por microSD, conectividad WIFI, etc.

Hace tan solo unos años, existían tres versiones de Netduino:

- **Netduino:** sin interfaz Ethernet, ni microSD.
- **Netduino Plus:** igual que Netduino pero con interfaz Ethernet y lector microSD.
- **Netduino Mini:** versión reducida compatible con Basic stamp2.

Estas tres versiones ejecutaban la versión .NET Microframework 4.1, e incluían un microcontrolador ARM® (concretamente el ARM7TDMI) funcionando a 48Mhz, contando con 512 KB de memoria Flash y 128 KB de RAM, a diferencia de los actuales Netduino 2, que se basan en los nuevos microcontroladores Cortex®.

La gran evolución de Netduino, sin duda, se lo debemos a su nuevo microcontrolador: un ARM® de 32 bits CPU Cortex® -M4 con FPU que funciona a una frecuencia de reloj hasta 168 MHz, consiguiendo un rendimiento de 210 DMIPS, contando además con un acelerador adaptable en tiempo real ART y una unidad de protección de memoria con DSP.

El ART Accelerator, combinado con la tecnología de 90 nm, permite 0 ciclos de espera en la ejecución del estado de la memoria Flash, proporcionando un rendimiento lineal hasta 168 MHz, desatando el rendimiento total del núcleo con un bajo consumo de energía, gracias a sus modos (Sleep, Stop, Standby) y al suministro de V_{BAT} para RTC.

El STM32 F4 serie ARM® Cortex®- M4 MCUs incluye 512 KB de memoria flash en el chip, 192 KB de SRAM, y 15 interfaces de comunicación que vamos a describir:
- Hasta 3 interfaces I²C.
- Hasta 4 USARTs / 2 UARTs (interface a10,5 Mbit/s ISO 7816, LIN control modem).
- Hasta 3 SPI (37, 5 Mbits / s), 2 con multiplexado full-duplex I²S.
- Interfaz SDIO.
- 2 × interfaces CAN (2.0b activo).

Además, el chip prevé conectividad avanzada por USB 2.0 / host / OTG y conectividad 10/100 Ethernet MAC, con DMA dedicado y soporte de hardware IEEE 1588v2, MII / RMII.

Entre el resto de características, destacan:

- Interfaz paralelo para LCD con modos 8080/6800.
- Dos convertidores de 12 bits D/A.
- DMA genérica: DMA 16 – stream controlador con FIFO y apoyo de ráfaga.
- Máximo de 17 temporizadores: hasta doce temporizadores de 16 bits y dos temporizadores de 32 bits de hasta 168 MHz, cada uno con hasta 4 IC / OC / PWM.
- Hasta 140 puertos de E/S con capacidad de interrupción.
- Verdadero generador de números aleatorios.
- Unidad de cálculo de CRC.
- RTC con precisión por debajo del segundo y calendario hardware.

2-Netduino 2

En la actualidad existen tres placas diferentes de **Netduino 2,** ejecutando todas las versiones .Net Microframework 4.2:

- **Netduino 2**: Es la versión más económica de Netduino, al no contar con interfaz Ethernet ni lector de microSD. Cuenta con un microcontrolador con el núcleo Cortex®-M3 de STMicro, concretamente el **STM32F2**, funcionando a una velocidad de reloj de 120Mhz, estando dotado de 192KB de espacio de almacenamiento para nuestro código y 60KB de RAM. Esta placa es compatible con otros escudos para Arduino, aunque, en algunos casos, se requerirán drivers NETMF. En el apartado de entradas salidas digitales, es muy generosa, contando con 22 GPIO, 6 PWM, 4 UART, I²C, SP y 6 canales ADC de 12 bits. Por último, **es posible añadirle almacenamiento extra** con un escudo con slot para microSD.
- **Netduino plus 2**: es un poderoso Netduino con **Ethernet integrado,** que además cuenta con una **ranura para microSD** en la misma tarjeta (soportando tarjetas de 2GB en adelante). Está basado en un microcontrolador Cortex®-M4 de STMicro, concretamente el STM32F4, funcionando a una velocidad de 168Mhz, contando con 384KB de espacio de almacenamiento para nuestro código y más de 100KB de RAM. La placa es también compatible con otros escudos para Arduino, pero se requerirán normalmente drivers NETMF. En el apartado de entradas y salidas digitales cuenta con 22 GPIO, 6 PWM, 4 UART, I²C, SP y 6 canales ADC de 12 bits.
- **Netduino Go**: es la versión modular de Netduino, basada en módulos Gobus, y constituye la opción ideal para los que no quieran soldar ningún cable. Cuenta con el microcontrolador Cortex®-M4 de STMicro, concretamente el STM32F4, (es decir, también el mismo de la plataforma Netduino plus 2), el cual funciona una velocidad de reloj de 168Mhz, contando con 384KB de espacio de almacenamiento para nuestro código, y más de 100KB de RAM. Según la filosofía de esta placa, tanto el interfaz Ethernet, como el interfaz SD, como los GPIO, PWM, UART, SPI, etc., se añaden aparte mediante módulos gobus.

Como vemos, la versión Netduino 2 cuenta **con cuatro veces más velocidad de reloj** (168 MHz) que su antecesor**, seis veces más espacio de código** (384KB), y el **doble de RAM** disponible (100 KB +).

Netduino 2 en Español

También viene con toda una serie de ricas nuevas características, incluyendo cuatro puertos serie, seis canales de PWM, y 12-bit ADC. Además la cabecera ICSP de 6 contactos, ha sido intercambiada por un encabezado Mini-JTAG de 10 pines.

En el apartado del software, Netduino 2 ejecuta NET Micro Framework, lo que significa dos cosas importantes: el código es **interpretado**, y hay un **recolector de basura**, ambas razones que sirven para hacer su código más fácil de codificar y, por supuesto más seguro.

Interpretado significa que usted escribe su código en C#, pero en lugar de ser compilado directamente a código ensamblador del ARM® (y posteriormente, a código de máquina), primero se compila en "**bytecode**", y el procesador ejecuta un intérprete para leer el código de bytes, en vez de ejecutar directamente el código máquina. Esto hará que la ejecución de éste sea un poco más lenta, pero como contrapartida, el código será más seguro, pudiendo detectar errores, como desbordamientos, accesos incorrectos a memoria, e incluso es posible verificar la información del tipo de objetos.

A propósito, código administrado o "**Managed code**", es un término acuñado por Microsoft que identifica el código fuente de un programa de ordenador, como es Microsoft ® Visual Studio 2010, que requiere para ejecutarse de forma obligatoria una máquina virtual CLR (Common Language Runtime), en nuestro caso .NET Framework.

Como aspecto diferenciador, la mayoría de los microcontroladores normalmente **no ejecutan código interpretado** hasta que se coloca en éstos **un intérprete**, como por ejemplo en Arduino que cuenta con un microcontrolador AVR y no utiliza intérprete, o en BASIC Stamp que cuenta con PIC y usa un sencillo intérprete de BASIC. A cambio Netduino, que incluye un microcontrolador STM32, como hemos visto, sí cuenta con un **potente intérprete CLR de c#.**

En la siguiente tabla, podemos ver la comparación de características de la versión actual de Netduino 2 Plus con la versión anterior de Netduino Plus:

	Netduino Plus	Netduino 2 Plus
Factor de forma	Arduino compatible	Arduino compatible
Medidas	iguales que Arduino	iguales que Arduino
Sistema operativo	.net Micro Framework 4.1	.net Micro Framework 4.2
PROCESADOR		

Velocidad	48 MHz	168 MHZ
Núcleo	ARM7TDMI	cortex-M4
Modelo	AT91SAM7X512	STM32F405RG

MEMORIA FLASH

Total	512KB	1MB
Usuario	128KB	384MB

RAM

Total	128KB	100+ kb
Usuario libre	28 Kb	100kb

I/O

Digital I/O	20	22
Interrupciones	20	22
Entradas analógicas	6X10-BIT	6x 12bits
Referencias	Rev.a:Externa (aref) Rev.B Seleccionable	3,3V en placa
PWM	6x16 bit	6x16bit

INTERFACES

USB 2.0	1XFULL SPEED	1XFULL SPEED
serie (UART)	2xTTL	2xTTL
ETHERNET	no	si
SPI	1x	1x
I2C	1x	1x

PERIFÉRICOS INTEGRADOS

SLOT SD	no	no
LED	1x	1x
PULSADOR	1x	1x

PROGRAMACIÓN

INTERFACE DEPURACIÓN	usb,serie	usb,serie
USUARIO BORRABLE	app,firmware,full	app,firmware,full

ALIMENTACIÓN

TENSIÓN ENTRADA	7.5-12 DC 5v (alimentado con USB)	7.5-12 DC 5v (alimentado con USB)
TENSIÓN SALIDA	3,3v -5V	3,3v -5V
TENSIÓN E/S		
CORRIENTE SALIDA E/S	2/8/16 mA	2/8/16 mA

CONSUMO ACTIVO	80mA	80mA
CONSUMO EN REPOSO	53mA	53mA
CONDICIONES AMBIENTE		
TEMP.OPERACIÓN	0..70°C	0..70°C
RoHS	si	si

En este libro, puesto que la idea es usar la conectividad proporcionada por el interfaz Ethernet, en todos los ejemplos usaremos la versión **Netduino 2 Plus,** que es la versión más potente hasta la fecha, aunque realmente no se requerirá la conectividad a la red hasta los últimos capítulos, donde realmente se explotan sus grandes posibilidades. No obstante, para todas aquellas personas que cuenten con la versión anterior, en el anexo 1, se dan las pistas de forma detallada para actualizar su Netduino + a la nueva versión Netduino 2+.

Una peculiaridad de la placa de Netduino 2 Plus, es que los pines del 1 al 14 pueden configurarse como entradas o salidas binarias digitales, pero para usarlas para otros fines, debe respetarse la siguiente configuración:

- Pines digitales 0-1: UART 1 RX, TX.
- Pines digitales 2-3: UART 2 RX, TX.
- Pines digitales 5-6: PWM, PWM.
- Pines digitales 7-8: UART 2 RTS, CTS.
- Pines digitales 9-10: PWM, PWM.
- Pines digitales 11-13: SPI MOSI, MISO, SPCK.

Respecto al resto de pines, correspondientes a las seis pines analógicos, éstas son las principales características:

- Pueden configurarse también como entradas o salidas binarias digitales, lo que sumadas a 14 procedentes de los pines 1 al 13, componen en total 20 entradas/salidas digitales.
- Pueden configurarse como entrada o salidas analógicas gracias al GPIO.
- Los pines 4 y 5 analógicos, además pueden configurarse como: I2C, SDA o SCL.

Resumiendo este apartado, las características más interesantes de la versión de Netduino 2 Plus, que se recomienda para el seguimiento de este libro, son las siguientes:

- *Microprocesador Cortex®-M4 de 32 bits funcionando a una velocidad de reloj de 168 MHz.*
- *384KB para almacenamiento de código.*
- *100 Kb+ de RAM.*
- *Soporte para tarjetas microSD de hasta 2 GB en la propia placa.*
- *Las salidas digitales pueden manejar hasta 25mA de corriente (perfectos para soportar directamente un led).*
- *Es compatible con Arduino "R3" para apoyar escudos futuros (además de escudos existentes, gracias a la tolerancia de 5 volt. digital de E / S de Netduino™ Plus 2).*
- *OneWire y Time Server directamente dentro del firmware NETMF.*
- *Espacio en flash para futuras características.*
- *Voltaje de entrada: 7,5 a 9,0 volt o con alimentación USB.*
- *25 mA. de corriente máxima en cada pin.*
- *Los pines digitales de I / O son de 3.3V pero tolerantes a 5v.*
- *Cuatro puertos serie.*
- *Seis canales de PWM.*
- *12-bit ADC.*

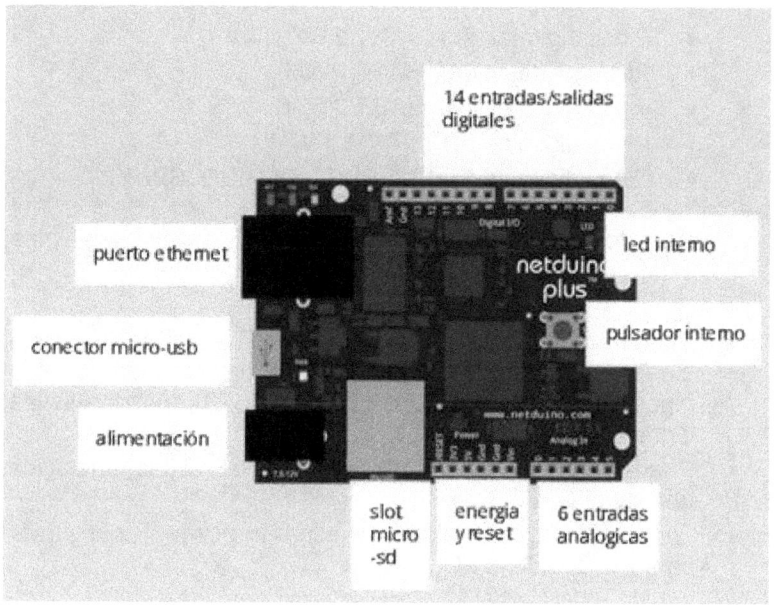

3-Preparación del entorno en Microsoft ® Windows 7

En el caso de disponer de un Netduino 2, Netduino plus 2 o cualquier placa anterior actualizada a .Net Micro Framework 4.2, y tener instalado Microsoft ® Windows 7 o Microsoft ® Windows XP en su ordenador, puede seguir el siguiente orden de instalación si desea que Microsoft ® Visual Studio 2010 sea su ambiente de desarrollo:

1. Instale Microsoft ® Visual c# Express 2010 desde la url siguiente: http://www.visualstudio.com/es-es#2010-Visual-CS
2. Instale el paquete gratuito. NET Micro Framework SDK v4.2 desde aquí:
   http://www.Netduino.com/downloads/MicroFrameworkSDK_NET MF42_QFE2.msi
3. Instale por último el SDK para la versión 4.2, en función de sistema operativo que tenga instalado en su ordenador (de 32bits o de 64bits):
   - S.O. de 32-bit ,V4.2.2.0 SDK Netduino:
     http://www.Netduino.com/downloads/Netduinosdk_32bit_NETMF42.exe
   - S.O. de 64-bit, V4.2.2.0 SDK Netduino:
     http://www.Netduino.com/downloads/Netduinosdk_64bit_NETMF 42.exe

Si dispusiera de una placa antigua Netduino, o Netduino Plus sin actualizar (lo cual no es lo más adecuado), puede seguir el siguiente orden en la instalación:

1. Instale Microsoft® Visual c# Express 2010 desde:
   http://www.microsoft.com/visualstudio/eng/downloads#d-2010-express
2. Instale NET Micro Framework SDK v4.1 desde la url:
   http://www.Netduino.com/downloads/MicroFrameworkSDK.msi
3. Instale por último el SDK Netduino para la versión 4.1 en función del sistema operativo de su ordenador:
   - S.O. de 32-bit, V4.1.0 SDK Netduino:
     http://www.Netduino.com/downloads/Netduinosdk_32bit.exe
   - S.O. de 64-bit, V4.1.0 SDK Netduino :
     http://www.Netduino.com/downloads/Netduinosdk_64bit.exe

NOTA: La información para preparar la plataforma Netduino en su PC está disponible en http://Netduino.com/downloads/

4-Preparación entorno en Microsoft ® Windows 8

En este apartado, veremos cómo es posible instalar este nuevo entorno de Microsoft ® en Netduino 2 Plus sobre el Sistema Operativo Microsoft ® Windows 8.

En la página oficial de Netduino, no hay referencias explicitas de la

instalación de Microsoft ® Visual Studio sobre el sistema operativo Microsoft ® Windows 8.

Para empezar, puede usar la versión profesional de Microsoft ® Visual Studio 2012 o bien la versión gratuita Microsoft ® Visual Studio Express 2012, ambas para escritorio.

Si decide instalar la versión gratuita deberá descargar Microsoft ® Visual Studio Express 2012 for Windows Desktop, que puede descargar en el siguiente enlace: http://www.microsoft.com/en-us/download/details.aspx?id=34673

Puede que el enlace de descarga no le funcione con Google® Chrome, por lo que sí es el caso, por favor pruebe el enlace anterior desde Microsoft® Internet Explorer.

La razón de aconsejar la instalación de la versión Microsoft ® Visual Studio 2012, se debe a que posteriormente deberemos instalar .NET Framework SDK v4.3 así como el SDK de Netduino v 4.3 y ambas versiones son compatibles precisamente con esa versión del producto, tanto para la versión profesional como la versión gratuita.

Una vez estemos en la citada página, seleccionaremos el idioma y pulsaremos "Download".

Microsoft ® nos ofrece la opción de descargar la imagen ISO para crear

un disco de instalación o bien hacer la instalación directamente desde la red.

Si decidimos realizar la instalación por red, chequearemos el ítem "wdexpress_full.exe" y pulsaremos "Next".

A los segundos, tendremos que aceptar la ejecución del instalador, e inmediatamente, empezará a descargarse el programa en sí, aunque previamente obtendremos la pantalla de aceptación de la descarga desde la propia web.

A los pocos minutos, empezará la instalación del programa informándonos del espacio en disco necesario para la instalación del producto. Si estamos de acuerdo, aceptaremos la licencia, y acto seguido, pulsaremos el botón **Instalar**.

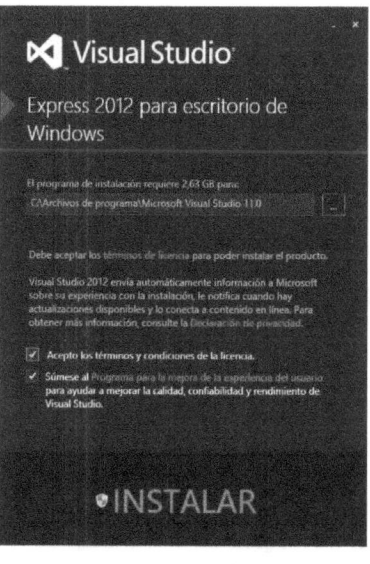

Ahora deberá esperar un rato indeterminado, que dependerá del ordenador y la velocidad de la red.

Una vez instalado el producto, al iniciar Visual Studio, si pensamos utilizar este entorno durante más de 30 días, tendremos que registrarnos online, o bien, dejarlo para más adelante, pues tendremos 30 días para hacerlo.

En caso de registrarnos, sólo deberemos introducir la cuenta de correo (preferiblemente de Microsoft ®) y un par de cuestiones, y nos ofrecerán

la clave del producto, la cual introduciremos en la caja anterior, y desde ese momento, ya tendríamos el entorno de Microsoft ® Visual Studio 2012 disponible sin restricciones.

En otras versiones de Microsoft ® Visual Studio, descargaríamos e instalaríamos el *SDK Netduino™ v4.2.1.0* (32 bits o 64 bits según la versión de Microsoft® Windows 8), pero en nuestro **caso podemos instalar el SDK v4.3**, para lo cual, una vez instalado Microsoft ® Visual Studio 2012 correctamente, se puede descargar e instalar la última versión *.NET Micro Framework SDK 4.3,* desde el siguiente enlace:
http://www.netduino.com/downloads/MicroFrameworkSDK_NETMF43_QFE1.msi

Al pinchar en el link anterior, empezará la descarga del instalador que deberemos aceptar. La instalación comenzará con la típica pantalla de bienvenida, que deberemos simplemente avanzar pulsando el botón "**Next**".

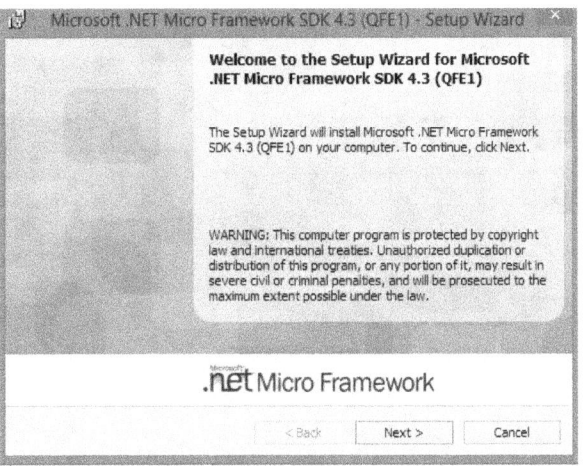

En caso de hacer la instalación por primera vez, pulsaremos en "**Typical**" y el botón "**Next**".

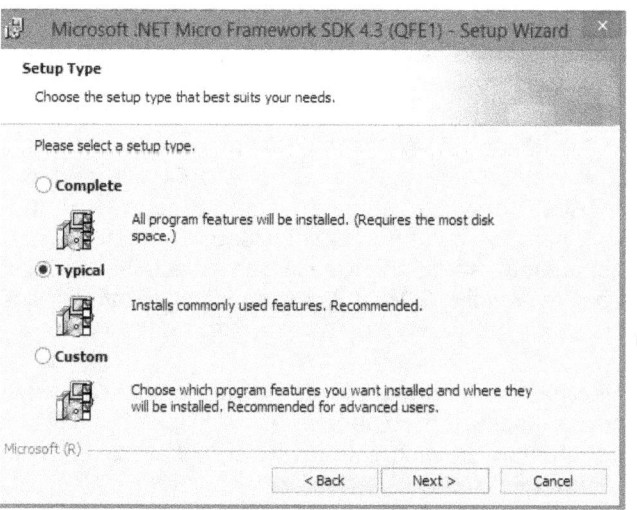

Ahora sólo queda copiar los ficheros, proceso que solo durará unos instantes.

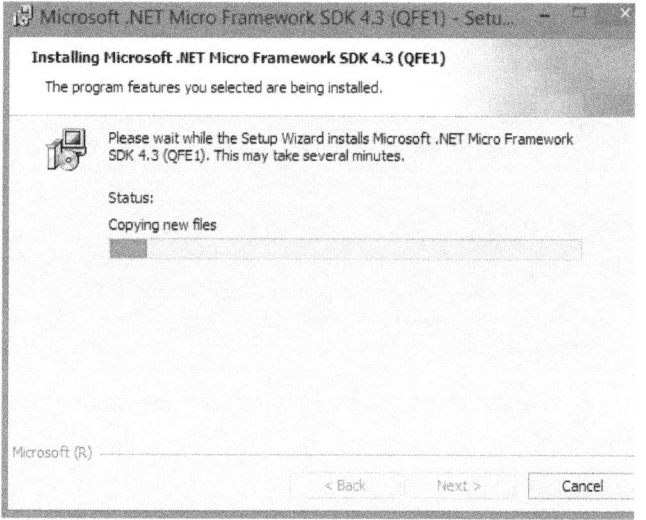

En unos segundos, ya habrá concluido la instalación del .Net Micro Framework, apareciendo el botón "**Finish**" para cerrar el instalador.

Netduino 2 en Español

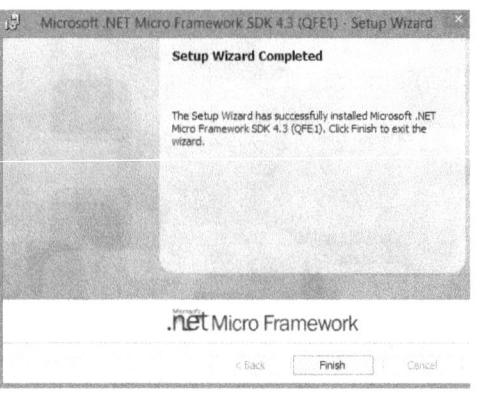

Finalmente para completar nuestro entorno de desarrollo, necesitamos también instalar la versión **NETMF 4.3**, lo cual instalará, tanto las plantillas necesarias para crear proyectos basados en las diferentes versiones de Netduino, como el driver adecuado para poder ser reconocida la placa por el sistema operativo.

Este es el enlace:
http://www.Netduino.com/downloads/Netduinosdk_NETMF43.exe

Una vez descargado el citado fichero y aceptemos la instalación, deberemos aceptar los términos de la licencia y pulsar "**Install**".

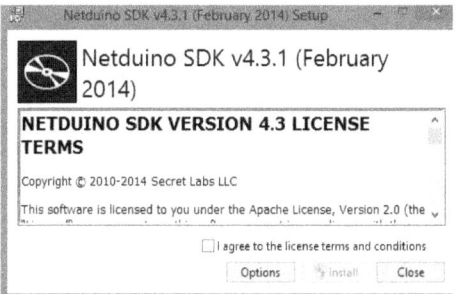

En unos segundos comenzará el proceso de instalación.

23

Carlos Rodríguez Navarro

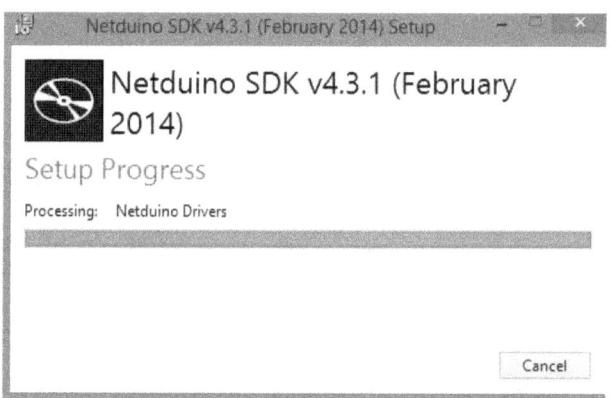

Mientras se instala el Netduino SDK v 4.3.1, puede que nos pida confirmación para la instalación del driver de Netduino 2, mensaje que obviamente aceptaremos pulsando "**Instalar**".

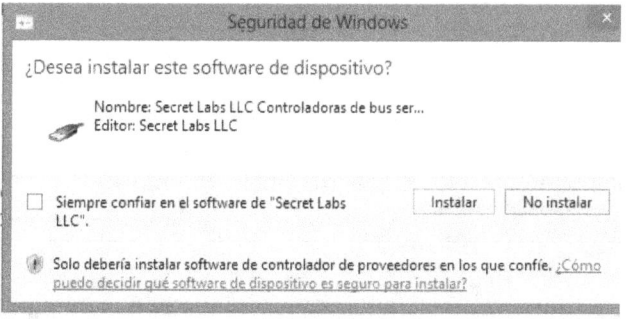

Tras unos segundos, concluirá la instalación del SDK, por lo que ya tendremos todo el entorno instalado preparado para empezar a probar los ejemplos que vamos a estudiar en este libro.

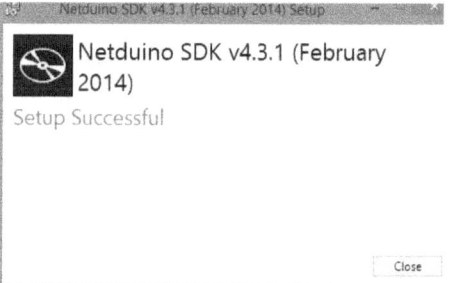

2 Ejemplo "Hola mundo"

1-Introducción

Una vez tenemos todas las herramientas instaladas en nuestro PC, para familiarizarnos con el entorno de desarrollo de Microsoft ® Visual Studio 2012, es muy interesante desarrollar nuestro primer programa "Hola Mundo".

Dado que Netduino no tiene una pantalla "estándar", es habitual que se use la conexión USB para depurar nuestros propios programas. Esto es posible de forma sencilla gracias a que **NETMF** provee al método **Print ("")**, dentro de la clase **Debug**, incluida en el espacio de nombres **Microsoft.SPOT**, es decir, un método que nos va a permitir obtener una impresión de lo que necesitemos en la ventana de Resultados del propio entorno de Visual Studio con fines de depuración para ver el desarrollo de nuestro programa.

Crear nuestro primer proyecto para Netduino, en realidad es muy sencillo: sólo ejecutaremos Microsoft ® Visual Studio 2012 y nos iremos a **Proyecto Nuevo,** lo cual nos llevara al asistente de proyectos de Visual Studio.

En la parte de Plantillas en Visual c#, nos iremos a Micro FrameWork, y desde ahí, seleccionaremos la plantilla en función de la versión de Netduino que dispongamos

.

Carlos Rodríguez Navarro

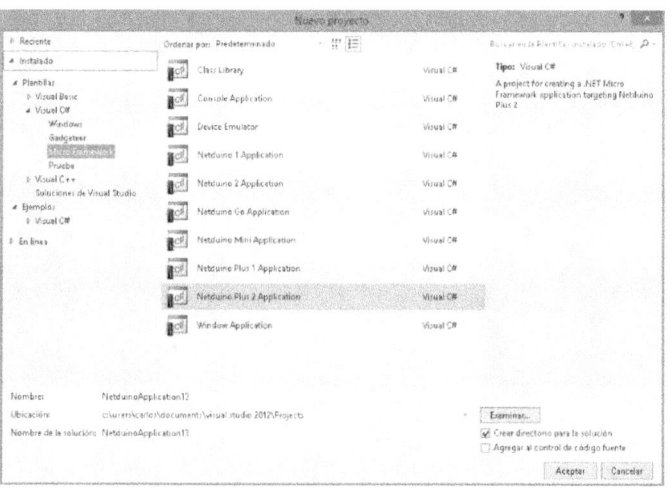

Una vez seleccionemos la opción de Netduino adecuada, se nos abrirá el explorador de Soluciones, con el esqueleto de la aplicación con el nombre predeterminado "NetduinoApplicationX".

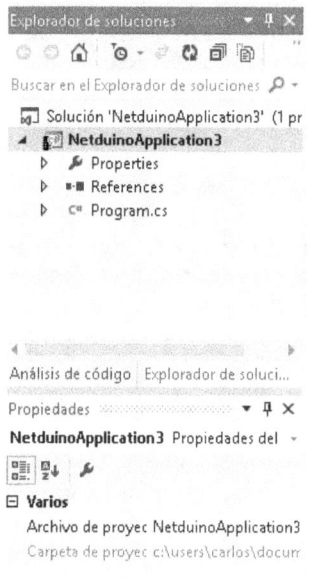

Antes de realizar ninguna acción, cambiaremos el nombre del proyecto con objeto de hacer más sencillo reconocerlo entre otros proyectos.

Netduino 2 en Español

Para ello, nos iremos dentro del árbol del proyecto en la rama Solución "NetduinoApplicactionX", y pulsaremos con el botón derecho "**Cambiar Nombre**" ,que en este caso, por ejemplo, lo llamaremos "Holamundo".

Asimismo, por coherencia, deberíamos cambiar también el nombre de la aplicación y el nombre de la clase, lo cual realizaremos pulsando con el botón derecho sobre cada nodo, dentro de la ventana del explorador de proyectos.

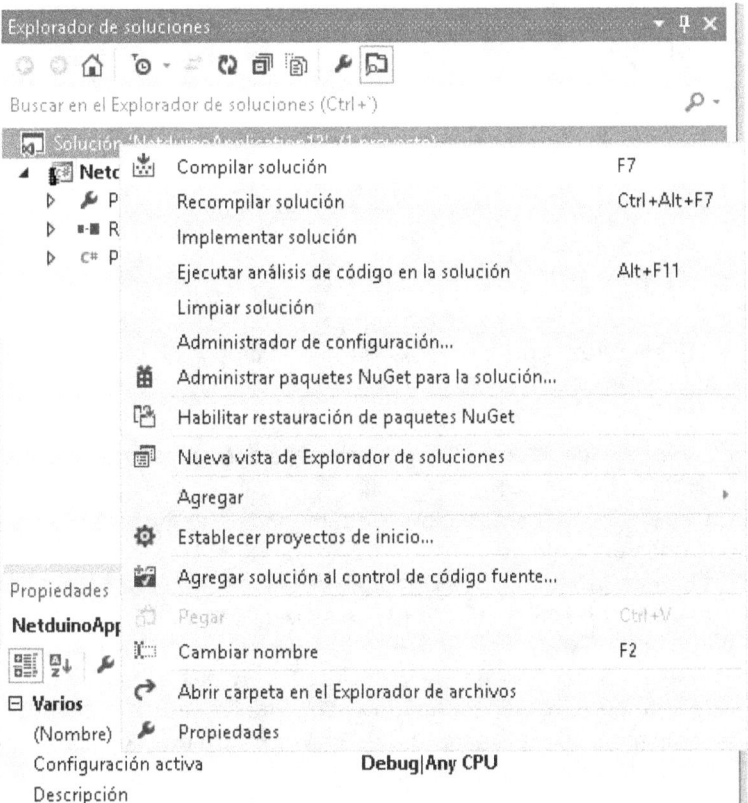

Una vez hayamos cambiado también el nombre en la aplicación (botón derecho "Cambiar nombre"), iremos a la opción **propiedades** y cambiaremos el **nombre del ensamblado** y el **Espacio de nombre** predeterminado, así como el **Objeto de Inicio** (este punto es de vital importancia, pues si no cambiamos este valor, no veremos acción alguna al ejecutar el programa).

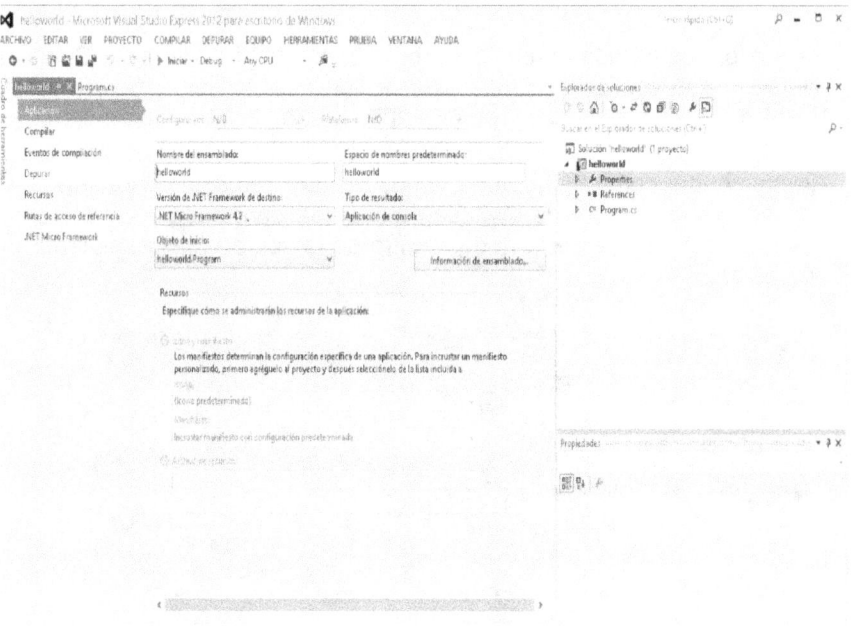

Respecto a las otras opciones, excepto la última, no es necesario modificar sus parámetros:

- La ventana de **Compilar**, podemos dejarlo tal como está por defecto.
- En eventos de compilación, podemos dejarlo también en las opciones por defecto.
- En cuanto la ventana de Depurar, debemos mantener la opción de **Configuración (Debug)** activa, y también la opción **plataforma (Any CPU)** activa.
- Respecto a la ventana de **Recursos** y **Rutas de acceso de referencia**, no es necesario tampoco cumplimentar nada.
- Por último, en la ventana .NET MicroFramework, son muy importantes las dos siguientes opciones:

  - En **Transport** seleccionar **USB**, para poder depurar nuestro programa por medio de una conexión USB hacia nuestra placa Netduino 2+ por medio del cable proporcionado (micro-USB a conector USB).
  - En **Device,** asegurarse que esta activado la versión Netduino que tenemos conectado.

Netduino 2 en Español

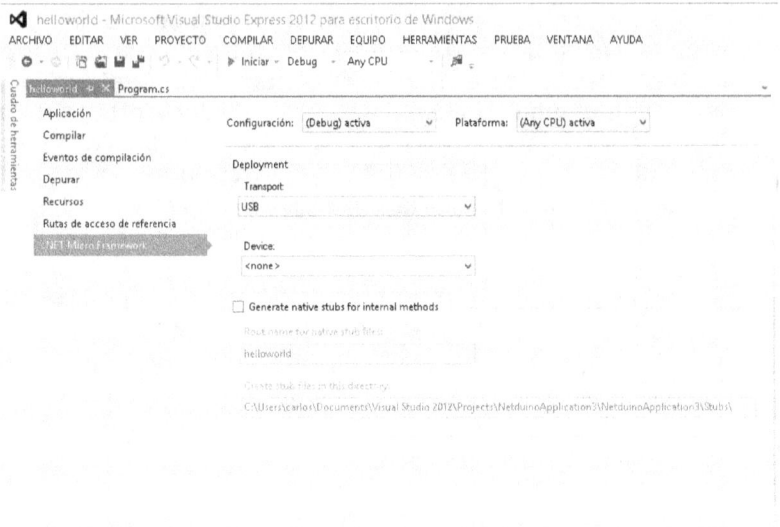

2-Creación de un nuevo proyecto

Teniendo el ambiente de desarrollo ya preparado, estamos en disposición de escribir el código en c#.net. Para ello, pulsaremos en la ventana de exploración de soluciones en la rama llamada **Program.cs** con doble clic.

NETMF, como código inicial, nos propone automáticamente el código que se muestra a continuación:

```csharp
using System;
using System.Net;
using System.Net.Sockets;
using System.Threading;
using Microsoft.SPOT;
using Microsoft.SPOT.Hardware;
using SecretLabs.NETMF.Hardware;
using SecretLabs.NETMF.Hardware.Netduino;

namespace NetduinoApplication4
{
public class HolaMundo
{
public static void Main ()
{
// write your code here
}
}
}
```

Como podemos ver, en el código de **programa.cs**, NETMF ya nos tiene preparado parte del código, introduciendo por nosotros los espacios de nombres más usuales (**System, System.Net, System.Net.Sockets, System.Threading, Microsoft.SPOT, Microsoft.SPOT.Hardware, SecretLabs.NETMF.Hardware** y **SecretLabs.NETMF.Hardware. Netduino**), al igual que el nombre de espacio de nombres del propio programa (**helloworld**), la clase principal (**Program**) y al menos la clase **Main.**

Las palabras reservadas public static void especifican el tipo de método (en este caso es **público,** visible a otras clases), **estático** (no necesita una instancia de la clase **HolaMundo** para ejecutar el método) y por último **void,** que indica que no devolverá ningún valor.

Como primer ejemplo, escribiremos en la salida definida por **Transport** la frase "Hola Mundo".

Para enviar dicha frase, utilizaremos simplemente la clase **Debug,** que pertenece al espacio de nombres **Microsoft.SPOT**, y el método *Print*, el cual escribe una salida de texto ASCII directamente al ambiente de desarrollo configurada en **Transport** (definida como USB, en la pestaña de .NET micro Framework, en la parte de conexiones , tal y como veíamos anteriormente).

Insertaremos, por tanto, el método Debug.**Print ("Hola Mundo")** dentro del cuerpo entre llaves {} de un bucle infinito que nunca terminará su ejecución (*puesto que su* condición nunca cambiará), mediante un bucle *while (true)*:

```
while (true)
{// write your code here
   Debug.Print ("Hola Mundo");
}
```

Finalmente, el código completo de nuestro primer programa para Netduino, es el que se muestra a continuación:

```
using System;
using System.Net;
using System.Net.Sockets;
using System.Threading;
using Microsoft.SPOT;
using Microsoft.SPOT.Hardware;
using SecretLabs.NETMF.Hardware;
using SecretLabs.NETMF.Hardware.Netduino;

namespace Netduino Application4
{
 public class HolaMundo
```

```
{
public static void Main ()
{
while (true)
{
// write your code here
Debug.Print ("Hola Mundo");
}
}
}
}
```

Para probar nuestro código, ya sólo nos falta pulsar el botón "**Iniciar**" (botón que hay debajo del menú superior que pone COMPILAR), y acto seguido, tras la compilación, empezará a ejecutarse en Netduino dicho código, que como hemos visto, se limita a sacar el saludo de forma ininterrumpida en la ventana "**Resultados**".

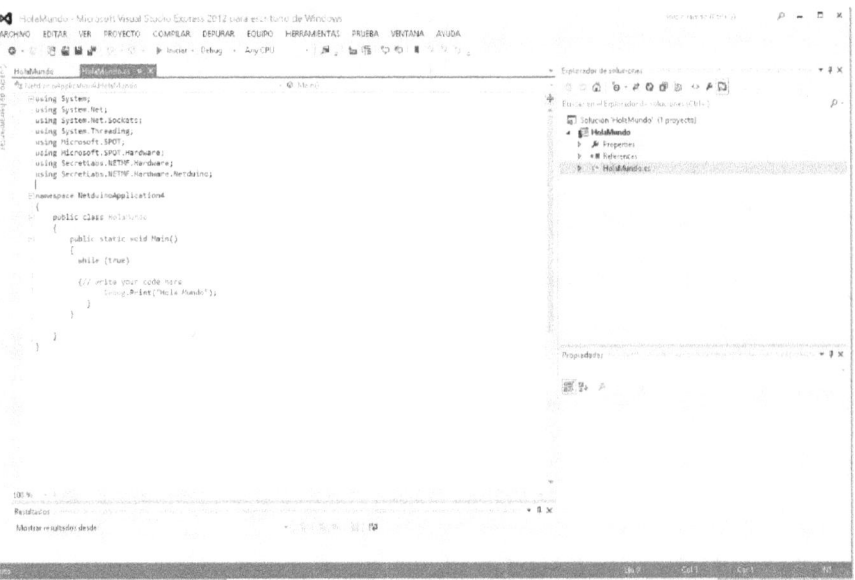

En caso de necesitar detener el programa, sólo habrá que pulsar el cuadradito rojo que hay debajo del menú **PRUEBA**, o si no reacciona, pulsar a la vez CONTROL+PAUSA.

Carlos Rodríguez Navarro

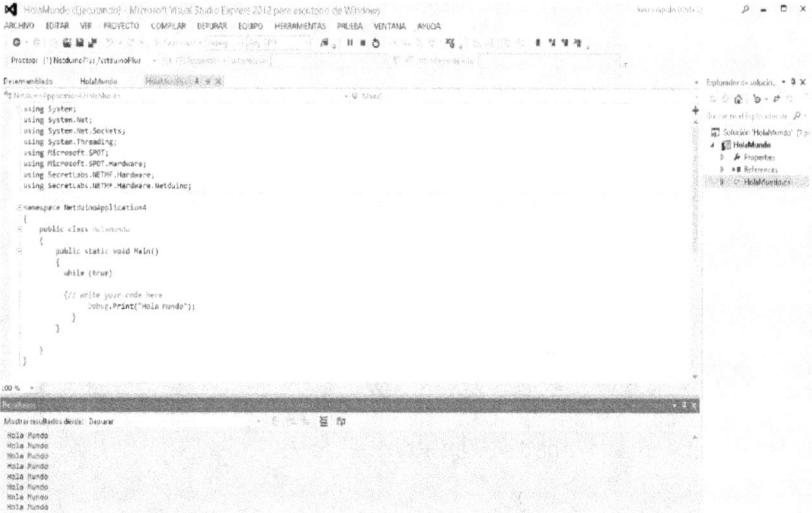

3 LED parpadeante

1-Introducción al uso de salidas binarias

Para empezar a usar el entorno de desarrollo de Microsoft ® Visual Studio, es muy interesante que el primer programa haya sido una salida simple por el puerto USB del mensaje "Hola Mundo", tal y como se mostró en el capitulo anterior, pero desgraciadamente, dado que ninguna de las placas de Netduino incluye "de serie" un display (tan sólo cuentan con un led interno y de un pequeño pulsador), es habitual que el primer proyecto real de Netduino, precisamente se centre en el manejo de este led interno, en lugar de sacar una salida por un puerto predefinido.

En el siguiente ejemplo que vamos a tratar, intentaremos pues manejar uno de los puertos internos de Netduino 2, al cual está conectado el led interno azul incluido en la propia placa.

En .net framework 4.2, para usar un pin físico como salida, se usa el objeto puerto de salida, lo cual es una instancia de la clase OutputPort, perteneciente al espacio de nombres Microsoft.SPOT.Hardware, y que por tanto se debe citar al principio del programa por medio de la clausula using.

Es importante destacar que de no incluir al espacio de nombres **Microsoft.SPOT.Hardware**, si quisiéramos usar este método en una clausula *using*, tendríamos que usar el nombre largo **Microsoft.SPOT.Hardware.OutputPort** en todas las veces donde fuera necesario usar la clase OutputPort.

Con más detalle, la clase *OutputPort* provee el método **Write,** el cual fuerza a un valor binario (0 o 1) al pin pasado como primer parámetro, según el valor del segundo parámetro, que se le pase como booleano (true o false) respectivamente.

En cuanto a la nomenclatura empleada para dirigirnos a los puertos digitales disponibles en Netduino 2, es la siguiente:

Hardware conectado	Uso	Constante
LED azul interno	Salida digital	Pins.ONBOARD_LED
PIN0	Entrada o salida digital	Pins.GPIO_PIN_D0
Pin1	Entrada o salida digital	Pins.GPIO_PIN_D1
Pin2	Entrada o salida digital	Pins.GPIO_PIN_D2
Pin3	Entrada o salida digital	Pins.GPIO_PIN_D3
Pin4	Entrada o salida digital	Pins.GPIO_PIN_D4
Pin5	Entrada o salida digital	Pins.GPIO_PIN_D5
Pin6	Entrada o salida digital	Pins.GPIO_PIN_D6
Pin7	Entrada o salida digital	Pins.GPIO_PIN_D7
Pin8	Entrada o salida digital	Pins.GPIO_PIN_D8
Pin9	Entrada o salida digital	Pins.GPIO_PIN_D9
Pin10	Entrada o salida digital	Pins.GPIO_PIN_D10
Pin11	Entrada o salida digital	Pins.GPIO_PIN_D11
Pin12	Entrada o salida digital	Pins.GPIO_PIN_D12
Pulsador interno	Entrada digital	Pins.ONBOARD_BTN
Pin0 analog.	Entrada o salida analógica	Pins.GPIO_PIN_A0 (*)
Pin1 analóg.	Entrada o salida analógica	Pins.GPIO_PIN_A1 (*)
Pin2 analóg	Entrada o salida analógica	Pins.GPIO_PIN_A2 (*)
Pin3 analóg.	Entrada o salida analógica	Pins.GPIO_PIN_A3 (*)
Pin4 analóg.	Entrada o salida analógica	Pins.GPIO_PIN_A4 (*)
Pin5 analóg.	Entrada o salida analógica	Pins.GPIO_PIN_A5 (*)

(*)Las entradas/salidas analógicas obviamente pueden emplearse como entradas o salidas **analógicas,** pero también pueden usarse como entradas o salidas **digitales** indistintamente.

2-Ejemplo de aplicación que emplea el led interno

En nuestro ejemplo definiremos la variable **ledPort,** como una instancia del objeto *OutputPort,* que hemos definido asociado al led interno mediante la constante Pins.ONBOARD_LED (ver tabla anterior), y con

estado apagado por defecto al inicializarlo.

```
var ledPort = new OutputPort (Pins.ONBOARD_LED, false);
```

Teniendo definido el objeto, para encender el led, basta utilizar el método **Write,** pasándole el valor **true** como único parámetro:

```
ledPort.Write (true);
```

Igualmente para apagar el led, bastará utilizar el método **Write,** pasándole el valor **false** como parámetro:

```
ledPort.Write (false);
```

En este ejemplo, también usamos el método **Sleep** perteneciente a la clase Thread, lo cual provocará una pausa del número de milisegundos que se le pase como parámetro (1500ms), lo cual en .net Framework es la mejor práctica, ya que permite llevar al hardware a un estado de bajo consumo para ahorrar energía.

```
Thread.Sleep (1500);
```

Respecto al método Debug.Print ("…"), como hemos visto en el capítulo anterior, lo empleamos en puntos del programa a afectos de depuración para corroborar las acciones que se hagan sobre el led interno.

Por último, insertaremos todo el código anterior dentro del cuerpo entre llaves{..} de un bucle *while (true),* cuya condición no cambiará, y por tanto, tampoco terminará su ejecución, por lo que en la práctica, esto permitirá que el led se encienda y se apague indefinidamente.

A continuación, se describe el código completo de un bucle infinito que enciende el led interno durante 1500ms, y después lo apaga también por 1500m, repitiendo el proceso continuamente:

```
using System;
using System.Net;
using System.Net.Sockets;
using System.Threading;
using Microsoft.SPOT;
using Microsoft.SPOT.Hardware;
using SecretLabs.NETMF.Hardware;
using SecretLabs.NETMF.Hardware.Netduino;

namespace EnciendeLed
{
public class EnciendeLed
    {
```

Carlos Rodríguez Navarro

```csharp
public static void Main ()
{
var ledPort = new OutputPort (Pins.ONBOARD_LED, false);
Debug.Print ("Led parpadeante");
while (true)
    {
            Debug.Print ("Led encendido");
            ledPort.Write (true); // enciende led placa
            Thread.Sleep (1500); // esperar 1500 ms
            Debug.Print ("Led Apagado");
            ledPort.Write (false); // apaga led placa
            Thread.Sleep (1500); // esperar 1500 ms
    }
    }
    }
}
```

Para probar nuestro nuevo código, sólo nos falta pulsar el botón "**Iniciar**" (botón que hay debajo del menú superior que pone COMPILAR), y acto seguido, tras la compilación, empezará a ejecutarse en Netduino 2, encendiendo y apagando el led interno durante 1500ms, y también sacando un texto por la ventana de depuración de "Resultados".

El resultado de la compilación lo puede ver en Anexo 3.

Y éste es el aspecto de la consola de Microsoft ® Visual Studio 2010 durante la ejecución de este programa:

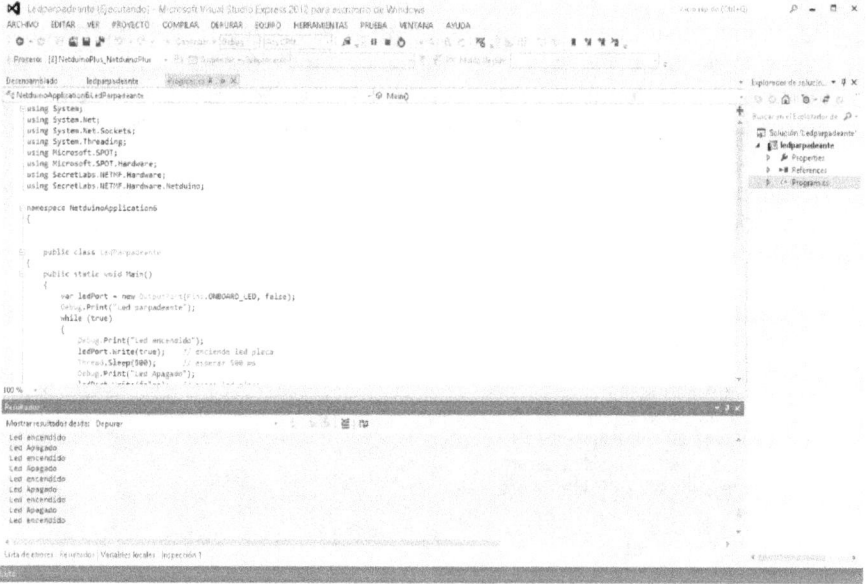

36

4 Entradas y Salidas digitales

1-Introducción a las entradas binarias

Tal y como ya adelantábamos en el capitulo anterior, con Netduino 2 no sólo es posible controlar salidas binarias, como por ejemplo el led interno azul cuyo ejemplo se describió, sino también es posible leer entradas digitales desde los mismos pines digitales configurados como entradas binarias.

La lectura de señales binarias lo haremos a través del objeto InputPort, el cual deberá ser instanciado en una variable, en nuestro ejemplo, en la variable "botón":

```
var boton = new InputPort (Pins.ONBOARD_BTN, false,
Port.ResistorMode.Disabled);
```

En el objeto InputPort, como vemos, se requieren tres parámetros:

*   **La entrada binaria:** este parámetro define el pin al que esté conectada la entrada física, como por ejemplo un sensor magnético, un interruptor, la salida de pulsos de vatímetro, etc. o en nuestro caso, por comodidad en este primer ejemplo, el propio pulsador interno. En Netduino 2 o Netduino con firmware .net 4.2 deberemos referenciarlas siguiendo la siguiente tabla:

PIN	PARÁMETRO
0	Pins.GPIO_PIN_D0

1	Pins.GPIO_PIN_D1
2	Pins.GPIO_PIN_D2
3	Pins.GPIO_PIN_D3
4	Pins.GPIO_PIN_D4
5	Pins.GPIO_PIN_D5
6	Pins.GPIO_PIN_D6
7	Pins.GPIO_PIN_D7
8	Pins.GPIO_PIN_D8
9	Pins.GPIO_PIN_D9
10	Pins.GPIO_PIN_D10
11	Pins.GPIO_PIN_D11
12	Pins.GPIO_PIN_D12
Pulsador interno	Pins.ONBOARD_BTN
0	Pins.GPIO_PIN_A0
1	Pins.GPIO_PIN_A1
2	Pins.GPIO_PIN_A2
3	Pins.GPIO_PIN_A3
4	Pins.GPIO_PIN_A4
5	Pins.GPIO_PIN_A5

Destacar, que dado que en el ejemplo usaremos el pulsador interno de la placa de Netduino 2 (presente en cualquiera de sus variantes), usaremos el parámetro **Pins.ONBOARD_BTN.**

- **Parámetro glitchFilter** (booleano): cuando leemos señales binarias, es muy habitual que ocurran "rebotes" o ruido, lo cual se traduce en que las lecturas no obtenemos un valor estable hasta que no nos hayamos cerciorado que el valor es el mismo en un determinado número de muestras. Si ponemos este parámetro al valor "**true**", forzaríamos a que Netduino 2 tome un número determinado de muestras al pulsar el botón hasta estar seguro que la lectura sea estable.

- **Parámetro resistormode** (booleano):poniéndolo a "**disabled**", indica que no se use la resistencia interna conectada al pin digital del micro-controlador.

Una vez instanciado en una variable el objeto InputPort, la forma de recuperar el valor de esa entrada binaria, es mediante el método **Read,**

el cual se encargará de almacenar en una variable booleana el valor binario obtenido.

Como vemos en el ejemplo, esta variable es "estadoboton", que es necesaria definirla anteriormente como booleana y con un valor por defecto **false**:

```
bool estadoboton=false;
```

2-Ejemplo de aplicación que usa el pulsador interno

Hemos visto como para usar el pulsador interno de Netduino 2, hemos de definir la variable botón, como una instancia del objeto *InputPort,* que hemos definido asociado al pulsador interno mediante la constante Pins.ONBOARD_LED (ver tabla anterior), sin filtro anti-rebotes y sin resistencia interna siguiendo la sintaxis siguiente:

```
var boton = new InputPort (Pins.ONBOARD_BTN, false,
Port.ResistorMode.Disabled);
```

Y posteriormente para leer el estado binario usando la expresión:
```
estadoboton=boton.Read ();
```

Asimismo, para usar el led interno, también definimos la variable **ledPort,** como una instancia del objeto OutputPort y que asociaremos al led interno con estado apagado por defecto al inicializarlo:

```
var ledPort = new OutputPort (Pins.ONBOARD_LED, false);
```

En este ejemplo, también usamos el método **Sleep** perteneciente a la clase Thread, la cual provocara una pausa del número de milisegundos que se le pase como parámetro (550ms), lo cual nos permitirá apreciar con claridad cada vez que se pulse el botón:

```
Thread.Sleep (550);
```

Respecto al método Debug.Print ("…"), como hemos visto en el capitulo anterior, lo empleamos en puntos del programa a afectos de depuración, es decir, para corroborar las acciones que se hagan sobre el led interno.

Como deseamos que el parpadeo del led interno responda al estado del pulsador, esta labor la realizaremos mediante un típico bloque de decisión if (estadoboton) {acción1} else {…accion2...}, donde la acción1 se realizará si el estado del pulsador es "**true**", (pulsador presionado) y la acción 2 en caso contrario:

```csharp
if (estadoboton)
{
Debug.Print ("bool estadoboton=false;");
ledPort.Write (true); // enciende led placa
Thread.Sleep (550); // esperar 550 ms
}
else
{
// Debug.Print ("Led Apagado");
ledPort.Write (false); // apaga led placa
}
```

Por último, insertaremos todo el código dentro de un bucle while (true), cuya condición no cambiará nunca, lo cual en la práctica permitirá crear un bucle infinito de comprobación del estado del pulsador y sus consecuentes acciones.

A continuación el código completo del ejemplo:

```csharp
using System;
using System.Net;
using System.Net.Sockets;
using System.Threading;
using Microsoft.SPOT;
using Microsoft.SPOT.Hardware;
using SecretLabs.NETMF.Hardware;
using SecretLabs.NETMF.Hardware.Netduino;

namespace EncindeLed
{
        public class EnciendLed
        {
                public static void Main ()
                {
                var ledPort = new OutputPort (Pins.ONBOARD_LED, false);

                var boton = new InputPort (Pins.ONBOARD_BTN, false,
                Port.ResistorMode.Disabled);

                bool estadoboton=false;

                Debug.Print ("Led parpadeante");
                        while (true)
                        {
                        estadoboton=boton.Read ();

                                if (estadoboton)
                                {
                                Debug.Print ("bool estadoboton=false;");
                                ledPort.Write (true); // enciende led placa
                                Thread.Sleep (550); // esperar 550 ms
                                }
                                else
                                {
```

Netduino 2 en Español

```
                                        // Debug.Print ("Led Apagado");
                                        ledPort.Write (false); // apaga led placa
                                        }
                          }
                    }
              }
}
```

Para probar nuestro nuevo código, sólo nos falta pulsar el botón "**Iniciar**" (botón que hay debajo del menú superior que pone COMPILAR), y acto seguido tras la compilación, empezará a ejecutarse en Netduino 2, encendiendo el led interno durante 550ms cada vez que se accione el pulsador interno, y también sacando un texto por la ventana de depuración de "Resultados".

5 Modulación por ancho de pulso

1-Introducción al PWM

Pulse Width Modulation (PWM), en español "**Modulación por Ancho de Pulso**", es una técnica consistente en alternar, a un frecuencia variable definida por la lógica, una señal entre dos valores extremos (en Netduino 2 entre 0 V y 3,3 V), asemejándose la salida en su forma más básica, a una onda cuadrada, donde tanto los valores mínimos como máximos tienen la misma duración, lo cual lo hace ideal para controlar servos, controlar el brillo de un led, generar sonidos o también como señal de reloj para otros circuitos.

En Netduino 2, como se comentó en el primer capítulo, se pueden usar cuatro de los puertos de salida (pines digitales 5, 6, 9 y 10), así como el puerto asociado al led interno, para generar señales de PWM.

Usar PWM en SDK 4.2 es ligeramente diferente a como se hacía en SDK 4.1, pero en todo caso, es bastante simple, como vamos a ver a continuación en el siguiente ejemplo, donde lo usaremos para controlar el brillo del led interno.

Antes de adentrarnos con el código, comentaremos algunos conceptos que requeriremos para entender la PWM:

- **Periodo**: Define el tiempo en microsegundos entre pico y pico de una señal periódica.

- **Frecuencia**: mide el número de repeticiones por unidad de tiempo, calculándose por tanto, como el valor recíproco del periodo.
- **Duración**: Define la duración del tiempo que la señal está a nivel uno lógico, para un ciclo en microsegundos. Este valor siempre deberá ser menor o como máximo igual que el periodo.
- **Ciclo de trabajo (DutyCycle)**: es la fracción de tiempo donde la señal es positiva (nivel lógico "1") y el periodo de la señal. Se calcula como la división entre la Duración y el Periodo, siendo del 0% si la señal siempre está a nivel bajo, y del 100% si la señal está permanente a nivel lógico alto.

En c# para definir una salida PWM, en lugar de utilizar el método OutputPort, lo haremos mediante la creación de una instancia de la clase PWM, que en SDK 4.2 tiene definidos 4 parámetros:

- El puerto de salida.
- La frecuencia en Hercios inicial.
- La frecuencia de trabajo Dutycycle.
- Inversión de la señal.

2-Ejemplo de aplicación con PWM

Para empezar, debemos definir una salida PWM, mediante la creación de una instancia de la clase PWM, que en SDK 4.2 tiene definidos 4 parámetros:

- El puerto de salida (en el ejemplo usaremos el led interno por lo que definiremos PWMChannels. PWM_ONBOARD_LED).
- La frecuencia en Hercios inicial (en el ejemplo se ha puesto de 100hz).
- La frecuencia de trabajo Dutycycle (en el ejemplo es del 50%).
- Inversión de la señal (en el ejemplo sin inversión).

De este modo, la definición del PWM para manejar el brillo del led interno será la siguiente:

```
PWM led1 = new PWM (PWMChannels.PWM_ONBOARD_LED, 100, .5, false);
```

Una vez definido la variable PWM, tenemos disponibles los siguientes métodos para manejar el brillo del led interno:

- Arrancar la generación de señal PWM mediante **led1.Start ()**.

Netduino 2 en Español

- Parar la generación de señal PWM por medio de **led1.Stop ()**.
- Ajustar la frecuencia en Hz mediante **led1.Frecuency (valor)**.
- Ajustar el ciclo de trabajo mediante **led1.DutyCycle (valor).**
- Ajustar el periodo mediante **led1.Period (valor).**
- Ajustar la duración mediante **led1.Period (duración).**

Ahora conocidos los métodos para gestionar PWM, para variar el brillo del led interno de un forma progresiva ascendente y luego descendente, (efecto ciclo "día-noche") lo haremos variando el ciclo de trabajo del citado led.

Para obtener una señal periódica ascendente y luego descendente, gracias a la potencia de Netduino y c#.net, emplearemos la función trigonométrica seno, para lo cual, usaremos como rango de datos valores comprendidos entre 4,712 y 10.995, valores a los que primero aplicaremos dicha función seno, después lo reduciremos a la mitad (multiplicando x 0.05), y luego los "elevaremos" (sumando al resultado 0.05).

Dicho proceso, para cada periodo, lo podemos ver gráficamente en las siguientes ilustraciones:

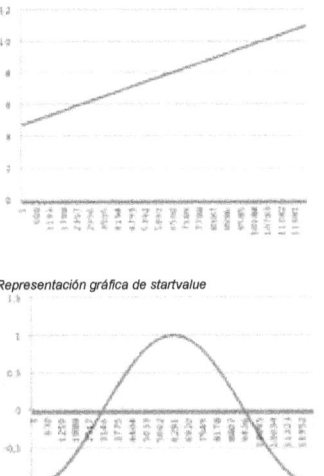

Representación gráfica de startvalue

Representación gráfica de System.Math.Sin (startValue)

45

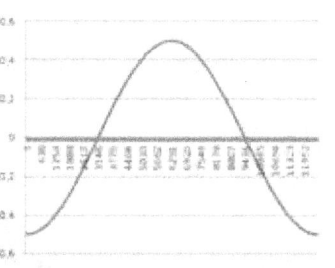

Representación gráfica de System.Math.Sin (startValue) * .5

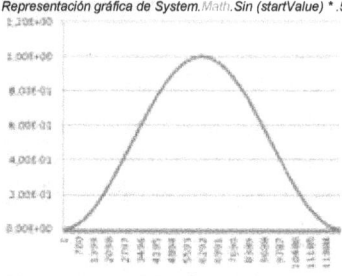

Representación gráfica de endvalue

Todo este proceso, lo haremos gracias a un bucle **for** que en c# se define con tres parámetros: la **condición de inicio** (en nuestro caso *startValue* = *4.712*), la **condición de repetición** (en nuestro caso siempre que *startValue* sea menos que 10,995), y por **último la acción** (en nuestro caso que *se incremente en 0.0005* la variable **startValue** cada vez que se repita el bucle).

Este sería el código resultante:

```
for (startValue = 4.712; startValue < 10.995; startValue = startValue + 0.0005)
{
......// calculo función campana
}
}
```

Dentro del citado bucle **for,** haremos el cálculo de la campana, que como se ha comentado, lo obtendremos aplicando la **función seno** a los valores que se obtienen del bucle **for,** después reduciendo a la mitad (multiplicando por 0.05), y luego elevando por encima del eje de abscisas (sumando al resultado 0.05).

```
endValue = System.Math.Sin (startValue) * .5 + .5;
```

Por último, cambiaremos dinámicamente el ciclo de trabajo del led interno, asignando el valor obtenido en la función anterior:

Netduino 2 en Español

```
led1.DutyCycle = endValue;
```

Para probar nuestro nuevo código, ya solo nos falta pulsar el botón
"**Iniciar**" (botón que hay debajo del menú superior que pone
COMPILAR), y acto seguido tras la compilación, empezara a ejecutarse
en Netduino 2**, encendiendo progresivamente** el led interno cada vez
que se mantiene pulsado el botón interno.

A continuación se describe el código completo de dicho ejemplo:

```csharp
using System;
using System.Net;
using System.Net.Sockets;
using System.Threading;
using Microsoft.SPOT;
using Microsoft.SPOT.Hardware;
using SecretLabs.NETMF.Hardware;
using SecretLabs.NETMF.Hardware.Netduino;

namespace Netduino Application15
{
        public class Program
            {
            public static void Main ()
                {
                double startValue, endValue;

                PWM         led1        =        new        PWM
                (PWMChannels.PWM_ONBOARD_LED, 100, .5, false); //
                asigna un PWM al led interno

                var boton = new InputPort (Pins.ONBOARD_BTN, false,
                Port.ResistorMode.Disabled); // usaremos el pulsador interno

                bool estadoboton = false;

                Debug.Print ("Led modulado por PWM");


                while (true)
                {
                estadoboton=boton.Read ();// lee estado del pulsador
                interno

                if (estadoboton) // si se pulsa el pulsador interno
                        {
                        led1.Start ();// enciende el led
                        Debug.Print ("Led Activado");

                        for (startValue = 4.712; startValue < 10.995;
                        startValue = startValue + 0.0005)
                            {
                            endValue       =       System.Math.Sin
                            (startValue) * .5 + .5; // calculo campana
```

47

Carlos Rodríguez Navarro

```
                        led1.DutyCycle = endValue; // cambia el
                    ciclo de trabajo del led placa
                        }
        }

        else // se NO se pulsa el pulsador interno
        {
        Debug.Print ("Led Apagado");
        led1.Stop (); // apaga led placa
        }



        }

    }

    }
}
```

6 Sensor de ruido

1-Procesamiento de señales analógicas

En este capítulo vamos a tratar como un simple sensor piezo-eléctrico se puede usar para detectar sonidos con Netduino 2, lo que nos permitirá usarlo como un sensor para golpes o "toques", por ejemplo, pegando el piezo- eléctrico a una superficie lisa.

Para conectar el sensor a Netduino 2, vamos a aprovechar la capacidad de leer señales analógicas por medio del CAD (convertidor analógico a digital), el cual permite leer el valor de un voltaje y "transformarlo" en un valor entre 0 y 1024, donde 0 representa 0 voltios, y 1024 representa 5 voltios en la entrada de cualquiera de de los 6 pines analógicos.

Las 6 entradas analógicas en Netduino se localizan en el conector inferior derecho marcadas con el texto "**Analog In**"(es decir entrada analógica).

La forma de referenciar las entradas analógicas en NETMF 4.2 se muestra en la tabla siguiente:

Hardware conectado	Uso	Constante
Pin0 analóg	Entrada analógica	Cpu.AnalogChannel.ANALOG_0
	salida analógica	

Pin1 analóg	Entrada analógica	Cpu.AnalogChannel.ANALOG_1
	Salida analógica	
Pin2 analóg	Entrada analógica o	Cpu.AnalogChannel.ANALOG_2
	salida analógica	
Pin3 analóg	Entrada analógica o	Cpu.AnalogChannel.ANALOG_3
	Salida analógica	
Pin4 analóg	Entrada analógica	Cpu.AnalogChannel.ANALOG_4
	Salida analógica	
Pin5 analóg	Entrada analógica o	Cpu.AnalogChannel.ANALOG_5
	Salida analógica	

Un piezo- eléctrico no es otra cosa que un dispositivo electrónico que suele usarse con bastante frecuencia para reproducir o detectar tonos. En el ejemplo que vamos a ver, hemos conectado el piezo- eléctrico en el pin de entrada analógica número 2 (ANALOG_2), lo cual nos permitirá leer un valor entre 0 V y 5 V, y no solamente HIGH o LOW (pines digitales) como en el caso de las entradas/salidas digitales.

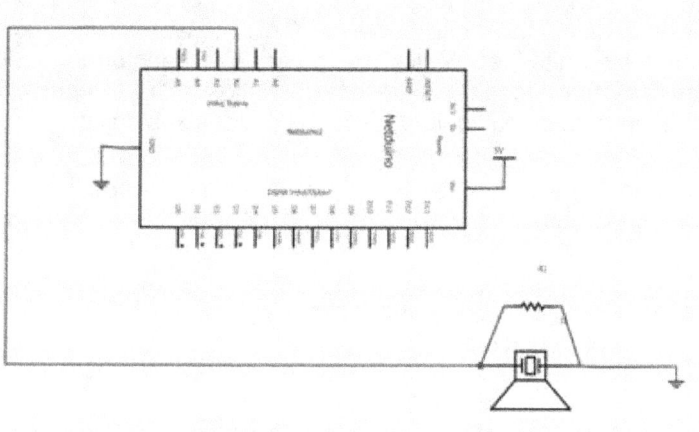

Un punto importante, es que los dispositivos piezo-eléctricos tienen polaridad, la cual, en los dispositivos comerciales, se indica

Netduino 2 en Español

normalmente con un cable rojo y uno negro para saber cómo conectarlo a la placa.

Cuando usemos estos dispositivos, debemos conectar el cable negro en la masa (0V) y el rojo en el pin de entrada, y además, deberemos conectar una resistencia de al menos 1 Mega-ohmio en paralelo al piezoeléctrico para mejorar la respuesta en frecuencia del conjunto (en el ejemplo se ha colocado directamente en los conectores del propio Netduino).

2-Ejemplo de aplicación que usa una señal analógica

En el siguiente código de ejemplo, capturáremos el ruido de un sensor piezoeléctrico, y si se sobrepasa cierto límite, se encenderá el led azul interno y sacaremos por la consola el mensaje "Se ha sobrepasado el umbral de sonido".

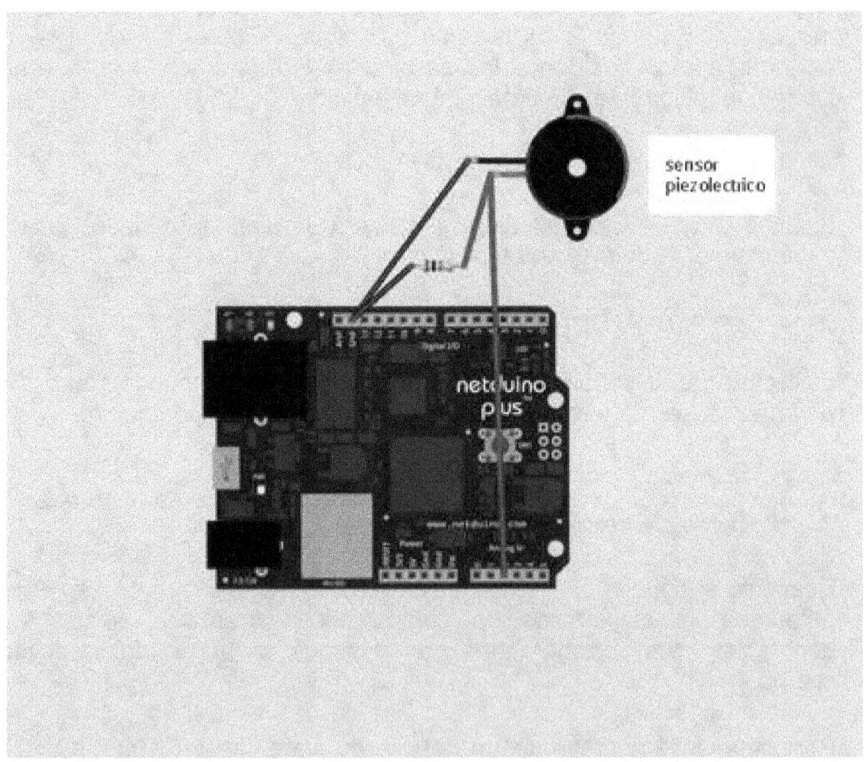

En el ejemplo, nuevamente definimos la variable **ledPort** como una instancia del objeto OutputPort,, que hemos definido asociado al led interno, y con estado apagado por defecto al inicializarlo.

```
var ledPort = new OutputPort (Pins.ONBOARD_LED, false);
```

Como aspecto importante, definimos la variable **voltajePorthumo** como una instancia del objeto AnalogInput, asociado al puerto 2 del conversor analógico-digital interno.

```
var voltagePorthumo = new Microsoft.SPOT.Hardware.AnalogInput
(Cpu.AnalogChannel.ANALOG_2); //sensor entrada analógica en A2 para
sensor de ruido
```

Teniendo definido el objeto, para encender el led, basta utilizar el método **Write** pasándole el valor **true** como único parámetro:

```
ledPort.Write (true);
```

En este ejemplo, también usamos el método **Sleep** perteneciente a la clase Thread, la cual provocara una pausa al menos del número de milisegundos que se le pase como parámetro (1500ms será suficiente para que apreciemos con el led encendido), indicando con esto que ha habido un ruido por encima del umbral definido.

```
Thread.Sleep (1500);
```

Igualmente para apagar el led, bastara utilizar el método **Write,** pasándole el valor false como parámetro:

```
ledPort.Write (false);
```

La lectura de la señal externa lo realizarnos con el método **Read** sin parámetros, salvando el resultado en una variable del tipo double:

```
double rawValue;
..............
rawValue = voltagePortRuido.Read ();
```

Respecto el método Debug.Print ("..."), como hemos visto en capítulos anteriores, lo empleamos en puntos del programa a afectos de depuración para corroborar las acciones que se hagan sobre el led interno.

Por último, insertaremos dentro del cuerpo entre llaves {..} de un bucle infinito que nunca terminara su ejecución, mediante un bucle while (true), cuya condición no cambiará, lo cual en la práctica permitirá que el led se encienda y se apague según la condición de ruido en un bucle infinito.

Para detectar cuando superamos un cierto umbral de ruido, utilizaremos un típico bloque condicional **if{ ..} else{..},** donde lo destacable es que al estar dentro del bucle infinito, se está comparando constantemente el valor obtenido en el conversor A/D con un valor de referencia que se ha estimado el 50% del tope. Si el lector lo desea, puede experimentar con otros valores.

```
if (rawValue >= tope) // si sobrepasa un umbral
{
  Debug.Print ("Se ha sobrepasado el umbral de sonido");
  ...
}
else
{
  ledPort.Write (false); // apaga led placa
}
```

Para probar nuestro nuevo código, ya solo nos falta pulsar el botón **"Iniciar"** (botón que hay debajo del menú superior que pone COMPILAR), y acto seguido tras la compilación empezará a ejecutarse en Netduino 2, encendiendo el led interno durante 1500ms y también sacando un texto por la ventana de depuración de "Resultados" cuando se detecta un ruido que sobrepasa cierto umbral o apagando éste en caso contrario.

A continuación el código completo de este ejemplo:

```
using System;
using System.Net;
using System.Net.Sockets;
using System.Threading;
using Microsoft.SPOT;
using Microsoft.SPOT.Hardware;
using SecretLabs.NETMF.Hardware;
using SecretLabs.NETMF.Hardware.Netduino;


namespace NetduinoApplication7
{
        public class Sensorsonido
        {
                public static void Main ()
                {
                  //Sensor de sonidos By CRN @ 2014

                  double tope = 0.5;
                  double rawValue;
                  string ttt;
```

Carlos Rodríguez Navarro

```csharp
var ledPort = new OutputPort (Pins.ONBOARD_LED, false);

var voltagePortRuido = new Microsoft.SPOT.Hardware.AnalogInput
(Cpu.AnalogChannel.ANALOG_2); //sensor entrada analógica en A2
para sensor de ruido


Debug.Print ("Leyendo puerto...");

            while (true) //bucle infinito
            {
            rawValue = voltagePortRuido.Read (); //captura valor
            sensor de humo
            ttt=rawValue.ToString ();
            Debug.Print ("valor =" + ttt);

                        if (rawValue >= tope) // si sobrepasa un
                        umbral
                        {
                        Debug.Print ("Se ha sobrepasado el
                        umbral de sonido");
                        ledPort.Write (true); // enciende led
                        placa
                        Thread.Sleep (1500);
                        }
                        else
                        {
                        ledPort.Write (false); // apaga led placa
                        }

            }
        }
}
```

7 Sensores de posición

1-Aplicación de los sensores de posición

Gracias a la evolución de la electrónica, y en particular a la capacidad de procesamiento, es posible llegar a soluciones que hace unos años hubieran sido impensables, como por ejemplo, en lugar de utilizar un conmutador mecánico, usar en su lugar una resistencia variable muy similar a un potenciómetro.

En una resistencia variable de este tipo, en función de la posición del cursor, variará su resistencia eléctrica entre este y un extremo, y con ello la diferencia de potencial en éstos. Si medimos este valor por medio de un convertidor A/D, se pueden realizar diferentes acciones, simplificando enormemente el cableado, pero como vamos a ver, haciendo más compleja la lógica.

En la figura siguiente podemos ver la gran sencillez de utilizar este tipo de sensores, pues simplemente requieren tres hilos: Vcc, Gnd y la salida que conectaremos a un puerto analógico para su conversión A/D.

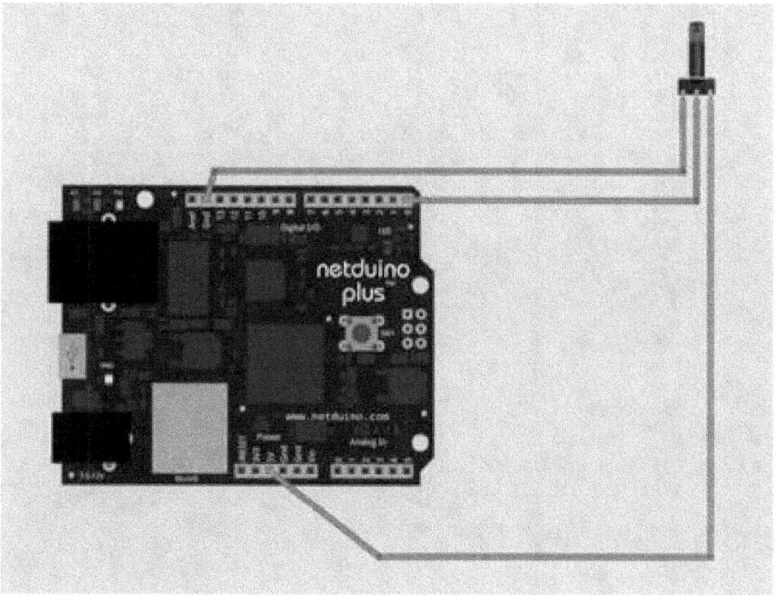

2-Ejemplo de aplicación que usa un sensor de posición

En el siguiente código de ejemplo, vamos a tomar el valor de un sensor de posición, de modo que si se llega al umbral máximo, encenderemos el led azul interno, así como también sacaremos por la consola el mensaje "Se ha llegado al tope".

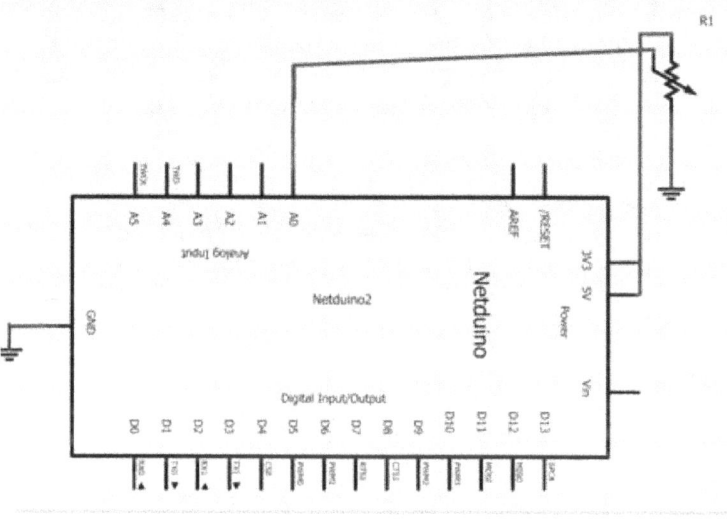

Netduino 2 en Español

En el ejemplo, nuevamente definimos la variable **voltagePortluz**, como una instancia del objeto OutputPort, que hemos definido asociado al led interno y con estado apagado por defecto al inicializarlo:

```
var ledPort = new OutputPort (Pins.ONBOARD_LED, false);
```

Como aspecto importante, definimos la variable **voltajePortluz**, como una instancia del objeto AnalogInput, que hemos definido asociado al puerto 0 del conversor analógico digital interno.

```
var voltagePotenciometro = new Microsoft.SPOT.Hardware.AnalogInput
(Cpu.AnalogChannel.ANALOG_0);
```

Teniendo definido el objeto, como se ha comentado en capítulos anteriores, para encender el led, basta utilizar el método **Write** pasándole el valor **true** como único parámetro:

```
ledPort.Write (true);
```

Y de modo similar, para apagar el led, bastará utilizar el método *Write*, pasándole el valor **false** como parámetro:

```
ledPort.Write (false);
```

La lectura de la señal externa, lo realizarnos con el método **Read** sin parámetros, salvando el resultado en una variable del tipo **double**:

```
double rawValue;
..............
rawValue = voltagePotenciometro.Read
```

Respecto el método Debug.Print ("..."), como hemos visto en otros capítulos, lo empleamos en puntos del programa a afectos de depuración para corroborar las acciones que se hagan sobre el led interno.

Por último, insertaremos dentro del cuerpo de un bucle infinito todo el código anterior mediante un bucle while (true), cuya condición no cambiar, lo cual en la práctica permitirá que el led se encienda y se apague, según la posición del sensor, en un bucle infinito.

Para detectar cuando superamos un cierto umbral máximo utilizaremos un bloque de decisión **if{ ..} else{..},** donde lo destacable es que al estar dentro del bucle infinito, se está comparando constantemente el valor obtenido en el conversar A/D con un valor de referencia que se ha estimado el 99.9% del tope (si el lector lo desea, experimente con otros

57

valores).

```
if (t >= tope) // si sobrepasa un umbral
{
Debug.Print ("Se ha llegado al tope");
ledPort.Write (true); // apaga led placa
}
else
{
ledPort.Write (false); // enciende led placa
}
```

Para terminar, dado que el ruido es un factor que cambia muchísimo, se ha optado por tomar 1000 muestras y luego obtener la media aritmética de estas (si el lector lo desea, experimente con otra cifra de muestras).

```
for (int aa = 0; aa < 1000; aa++) //1000lecturas
{
rawValue    =    voltagePotenciometro.Read    ();    //captura    valor
potenciómetro

media = rawValue * 100000;// lectura

t =media +t;


}

t /= 1000; //tomamos la media
```

Para probar nuestro nuevo código, ya solo nos falta pulsar el botón "**Iniciar**" (botón que hay debajo del menú superior que pone **COMPILAR**), y acto seguido tras la compilación, empezará a ejecutarse en Netduino 2, encendiendo el led interno sólo al llegar el sensor a la posición tope ,y también, sacando un texto por la ventana de depuración de "Resultados".

```
using System;
using System.Net;
using System.Net.Sockets;
using System.Threading;
using Microsoft.SPOT;
using Microsoft.SPOT.Hardware;
using SecretLabs.NETMF.Hardware;
using SecretLabs.NETMF.Hardware.Netduino;


namespace Netduino Application
{
```

Netduino 2 en Español

```csharp
public class LectorPotenciometro
{
        public static void Main ()
        {
        //Lectura de una resistencia variable By CRN @ 2014

        double tope =100050;
        double rawValue,media,t;
        string ttt;
        var ledPort = new OutputPort (Pins.ONBOARD_LED, false);
        var voltagePotenciometro = new
        Microsoft.SPOT.Hardware.AnalogInput
        (Cpu.AnalogChannel.ANALOG_0); //sensor entrada analógica en A2
        para sensor de ruido
        t = 0;

        Debug.Print ("Leyendo puerto...");

                while (true) //bucle infinito
                {
                        for (int aa = 0; aa < 1000; aa++) //1000lecturas
                        {
                         rawValue   =   voltagePotenciometro.Read    ();
                        //captura valor potenciómetro
                         media = rawValue * 100000;// lectura
                         t =media +t;
                        }


                t /= 1000; //tomamos la media
                ttt = t.ToString ();
                Debug.Print ("valor =" + ttt);

                        if (t >= tope) // si sobrepasa un umbral
                        {
                         Debug.Print ("Se ha llegado al tope");
                         ledPort.Write (true); // apaga led placa
                        }
                        else
                        {
                         ledPort.Write (false); // enciende led placa
                        }
                }

        }
    }
}
```

8 Sensor de Temperatura

1-Introducción

Para los proyectos con Netduino 2 en los que se requiera una forma compacta y rápida una sonda termométrica, tenemos a nuestra disposición el famoso sensor de Texas Instruments **LM35**.

Dicho sensor analógico, nos permite realizar medidas de temperatura de una forma bastante precisa, a través de las entradas analógicas de Netduino 2 (pines A0-A5), sin necesidad de emplear ninguna librería específica para su programación.

Gracias al principio físico de que al aumentar la temperatura de una unión pnp, lo hace también la caída de tensión entre la base y el emisor (el Vbe), amplificando el cambio de voltaje, es fácil generar una **señal analógica que sea directamente proporcional a la temperatura** (ha habido algunas mejoras en la técnica, pero, en esencia, ese sigue siendo el principio).

Este tipo de sensores no tienen partes móviles y son bastantes precisos, tienen un coste bajo, nunca se desgastan, no necesitan calibración, funcionan bajo muchas condiciones ambientales, ofrecen lecturas muy consistentes y además son muy fáciles de usar, como vamos a ver a continuación.

En este proyecto, usaremos el famoso LM35, el cual es un sensor de temperatura con una precisión calibrada de 1°C, con un rango de medición desde -55°C hasta 150°C, siendo la salida lineal (cada grado

centígrado equivale a 10mV). Este circuito integrado se encuentra en diferentes tipos de encapsulado, el más común es el TO-92, utilizado por transistores de baja potencia.

El LM35 no requiere de circuitos adicionales para calibrarlo externamente, dependiendo únicamente la tensión de salida de la tensión de alimentación, y por supuesto de la temperatura. La baja impedancia de salida, su salida lineal y su precisa calibración, hacen posible que este integrado sea instalado fácilmente en muchos circuitos de control. Además, debido a su baja corriente de alimentación, se produce un efecto de auto-calentamiento muy reducido que apenas afecta a su precisión.

Resumidamente, estas son sus algunas de sus características más relevantes:

- Está calibrado directamente en grados Celsius.
- La tensión de salida es proporcional a la temperatura.
- Tiene una precisión garantizada de 0.5°C a 25°C.
- Opera entre 4 y 30 voltios de alimentación.
- Baja impedancia de salida.
- Baja corriente de alimentación (60uA).
- Bajo costo.

Para empezar a usar el LM35, basta con conectar la patilla izquierda a la energía (desde 2.7 a 5.5V) y la patilla derecha a tierra. A continuación, el conector central tendrá una tensión analógica, que debería ser directamente proporcional (lineal) a la temperatura, siendo la tensión analógica independiente de la fuente de alimentación y proporcional a la temperatura ambiente.

Para procesar esta señal con Netduino 2, como tenemos una señal analógica, deberemos conectar ésta a un pin analógico de la placa, como son cualquiera de los pines A0 a A5 (en el ejemplo usaremos A0).

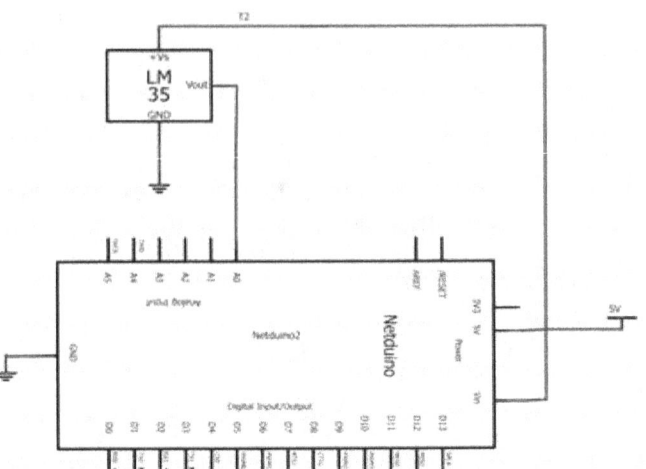

Para convertir el voltaje a la temperatura en grados centígrados, sólo tiene que utilizar la siguiente fórmula básica:

$$\text{Temperatura en } ^\circ C = [(\text{lectura DAC}) *20] / 0.7 \quad (*)$$

(*)-Esta fórmula se obtiene tomando la lectura del DAC a la temperatura de 20ºC alimentando el LM35 con 5V).

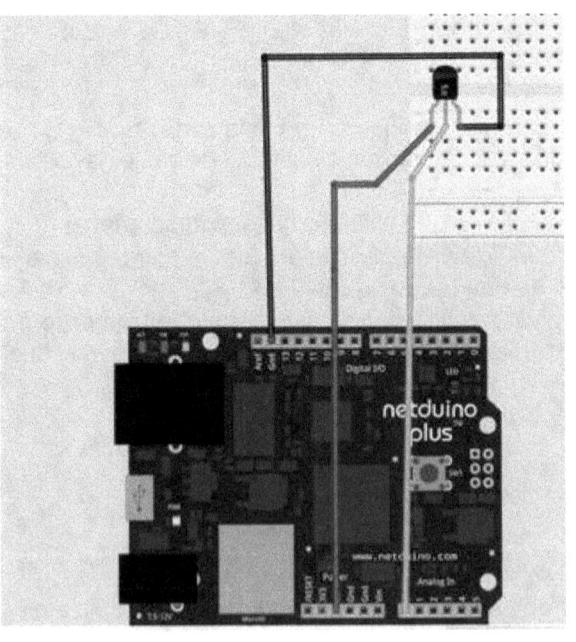

2-Ejemplo de aplicación

En el siguiente código de ejemplo, tomaremos un número variable de muestras de temperatura, obtendremos la media, y sacaremos por consola el valor en un bucle infinito, para lo cual en primer lugar, definimos la variable **voltajePort** como una instancia del objeto AnalogInput, que hemos definido asociado al puerto 0 del conversor analógico analógico-digital interno:

```
var voltagePort = new Microsoft.SPOT.Hardware.AnalogInput
(Cpu.AnalogChannel.ANALOG_0);
```

La lectura de la señal externa procedente del LM35, la realizaremos con el método **Read** sin parámetros, salvando el resultado en una variable del tipo **double**:

```
double rawValue;
..............
rawValue = voltagePort.Read ();
```

Con objeto de tomar una medida estable de la temperatura para calcular la temperatura real, tomaremos un número suficiente de muestras (en el ejemplo se ha calculado que 100 muestras es un número suficiente).

En cada muestra, después de obtener el valor de la conversión el DAC, lo convertiremos a grados centígrados mediante una función de conversión, y al finalizar la toma de las 100 muestras, calcularemos la media de todas las medidas.

Estas acciones, las haremos gracias a un bucle **for(a; b; c;){..}**, que en lenguaje c# se define de un forma similar al lenguaje java con tres parámetros:

   A. La condición de inicio (en nuestro caso **aa=0**).
   B. La condición de repetición (en nuestro caso siempre **que aa sea menor que 100**).
   C. La acción (en nuestro caso que **se incremente en una unidad** la variable **aa** cada vez que se repita el bucle).

Es decir, el bucle iría definido del siguiente modo:

```
for (int aa = 0; aa < 100; aa++) //100lecturas
{ rawValue = voltagePort.Read ();
.......
}
```

Dentro del bucle **for** una vez tomada la muestra, realizaremos la

Netduino 2 en Español

conversión a grados centígrados mediante la fórmula de conversión, es decir, multiplicando por 20 el valor obtenido por el conversor D/A, y luego dividiendo el valor obtenido por 0.07 (tenga cuidado con esta fórmula, pues será diferente si se alimenta el LM35 con otra tensión diferente de 5V cc):

```
voltage = rawValue*20;//
double temperatureC = voltage / 0.07336335665069986;
```

Calculada la temperatura de una muestra en grados centígrados, ya sólo queda ir sumando el valor en grados centígrados de todas las muestras, para luego dividir el resultado por el número de muestras (en nuestro caso 100), y obtener por último la media, que finalmente sacaremos por la ventana de depuración de "**Resultados**".

```
for (int aa = 0; aa < 100; aa++) //100lecturas
{
rawValue = voltagePort.Read ();
 voltage = rawValue*20;//
 double temperatureC = voltage / 0.07336335665069986;
 tt = temperatureC + tt;
}
tt /= 100; //tomamos la media
```

Respecto el método Debug.Print ("..."), como hemos visto en capítulos anteriores, lo emplearemos para sacar por consola el valor de la media del valor de temperatura obtenida en el cálculo anterior:

```
Debug.Print ("temperatura ºC=" + tt);
```

Por último, insertaremos todo el código anterior dentro del cuerpo de un bucle infinito while (true), cuya condición nunca cambiará, lo cual en la práctica permitirá que se tomen muestras de temperatura de forma ininterrumpida.

Para probar nuestro nuevo código, ya solo nos falta pulsar el botón "**Iniciar**" (botón que hay debajo del menú superior que pone COMPILAR), y acto seguido, tras la compilación, empezará a ejecutarse en Netduino 2 este pequeño programa que sacará la temperatura en grados centígrados de la sala de forma ininterrumpida por la ventana de depuración de "Resultados".

Este es el código completo del ejemplo:
```
using System;
using System.Net;
using System.Net.Sockets;
using System.Threading;
```

```csharp
using Microsoft.SPOT;
using Microsoft.SPOT.Hardware;
using SecretLabs.NETMF.Hardware;
using SecretLabs.NETMF.Hardware.Netduino;

namespace Netduino Application9
{
        public class Program
        {
                public static void Main ()
                {

                    var voltagePort = new Microsoft.SPOT.Hardware.AnalogInput
                    (Cpu.AnalogChannel.ANALOG_0);

                    double tt = 0;
                    double rawValue,voltage;

                        while (true) //se repite indefinidamente
                        {
                        for (int aa = 0; aa < 100; aa++) //100lecturas
                            {
                                rawValue = voltagePort.Read ();
                                voltage = rawValue*20;//
                                double    temperatureC    =    voltage    /
                                0.07336335665069986;
                                tt = temperatureC + tt;
                            }

                        tt /= 100; //tomamos la media
                        Debug.Print ("temperatura ºC=" + tt);
                        }
                }
        }
}
```

Notas:

1. La fórmula de conversión se basa en que alimentando el LM35 se obtiene el valor 0.07336335665069986 si la temperatura es 20º. Si lo desea puede utilizar otra función de conversión, por ejemplo tocando el sensor con un cubo de hielo, preferentemente en una bolsa de plástico para que no se moje con agua el circuito, y apuntando el valor obtenido del DAC.

2. El número de muestras se ha establecido a 100, por ser un compromiso entre un número no muy elevado a la vez que se obtiene una suficiente estabilidad en la lectura. Si el lector lo desea, puede experimentar con otros valores.

3- Sensores DHT11 y DHT22

Es habitual que en ambiente Arduino se usen profusamente sensores compuestos como el sensor DHT11, un dispositivo que incorpora un sensor de temperatura y humedad complejo constituido por dos sensores resistivos (NTC y sensor de humedad) y un micro-controlador de 8 bits integrado.

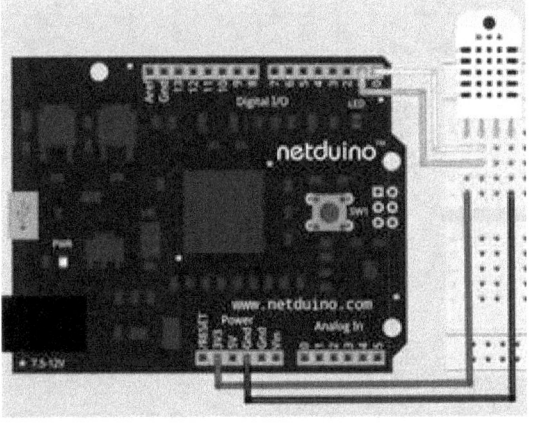

Nota: En el esquema se obvia la resistencia externa de 1K entre DO-D1 y +3,3V dado que con Netduino soporta por parámetro el uso de resistencia interna de pull-up.

El sensor dispone de una salida de señal única digital calibrada en laboratorio, lo cual asegura una alta calidad y una fiabilidad a lo largo del tiempo, pero en cambio esta característica, hace que sea difícil de gestionarlo en Netduino con NETMF 4.2 debido a que la resolución del temporizador del sistema no es suficiente para medir con precisión la duración de los impulsos del sensor, pudiendo suceder que la lectura del sensor falle debido a la suma de comprobación no válida, por lo general para valores específicos de humedad o temperatura.

Bien es cierto que se podía ignorar el error, o tratar valor BitThreshold, cambiar el firmware para mejorar la precisión del temporizador, etc. Pero, en todo caso, se precisan algoritmos más complejos para la lectura de la señal digital compuesta, así como dos líneas de datos digitales (normalmente D0 y D1) cuando, normalmente para una medida estándar de temperatura, sólo se requiere una entrada analógica con bastante poco lógica, como hemos visto en este capítulo.

En todo caso, para los interesados en usar cualquiera de estos sensores

compuestos con Netduino 2+, por favor consulte el lector el siguiente hilo sobre este tema en el foro oficial de Netduino http://forums.Netduino.com/index.php?showtopic=2560

9 Sensores de luz

1-Introducción a los LDR

Una fotorresistencia llamada también fotorresistor, fotoconductor, célula fotoeléctrica o resistor dependiente de la luz o simplemente **LDR** (en inglés light-dependent resistor), es un componente electrónico basado en el efecto fotoeléctrico, cuya resistencia es inversamente proporcional a la luz que le incide. Esto se traduce en que disminuye su resistencia con el aumento de intensidad de luz incidente, de modo que la resistencia eléctrica es baja cuando hay luz incidiendo en ella (puede descender entre 50 y 100 ohm) y es muy alta cuando está a oscuras (varios mega-ohmios).

Una típica LDR está hecha de un semiconductor de alta resistencia como el **sulfuro de cadmio** (CdS), de modo que si la luz que incide en el dispositivo es de alta frecuencia, los fotones son absorbidos por las elasticidades del semiconductor dando a los electrones la suficiente energía para saltar la banda de conducción. El electrón libre que resulta, y su hueco asociado, conducen la electricidad, de tal modo que **disminuye la resistencia,** es decir varia su resistencia según la cantidad de luz que incide en la célula, de modo que **cuanta más luz incide, más baja es la resistencia.**

Además de la luz visible, las células ldr también son también capaces de reaccionar ante una amplia gama de frecuencias, incluyendo el infrarrojo (IR), y la luz ultravioleta (UV), construyéndose de hecho con otros materiales como el Ge-Cu (que funciona dentro de la gama más baja o "radiación infrarroja").

Un aspecto importante respecto a las ldr, es que la variación del valor de la resistencia de estas tienen un cierto retardo (típico décima de segundo), que es diferente si se pasa de oscuro ha iluminado, o de iluminado a oscuro, lo cual limita el uso de las LDR en aplicaciones en las que la señal luminosa varíe con rapidez. No obstante, esta lentitud da ventaja en algunas otras aplicaciones donde la lentitud de la detección no es importante, ya que se filtran variaciones rápidas de iluminación que podrían hacer inestable un sistema (por ejemplo para saber si es de día o es de noche).

Estas células se fabrican en diversos tipos y formas, pudiéndose encontrar en muchos artículos de consumo, como por ejemplo en cámaras digitales, medidores de luz, relojes con radio, alarmas de seguridad o sistemas de encendido y apagado del alumbrado de calles.

Como toma de contacto de medición sencilla y muy asequible, vamos a tratar en este ejemplo la adquisición del nivel de luz medido con una sencilla LDR, por medio de un puerto analógico de Netduino 2.

Para conectar una LDR a una placa Netduino 2, simplemente haremos un pequeño divisor de tensión formado por la LDR y una resistencia de 220 ohmios. Los extremos de este divisor lo conectaremos respectivamente a 5V y a masa (0V), conectando el punto central a una entrada analógica (en el ejemplo el pin A1).

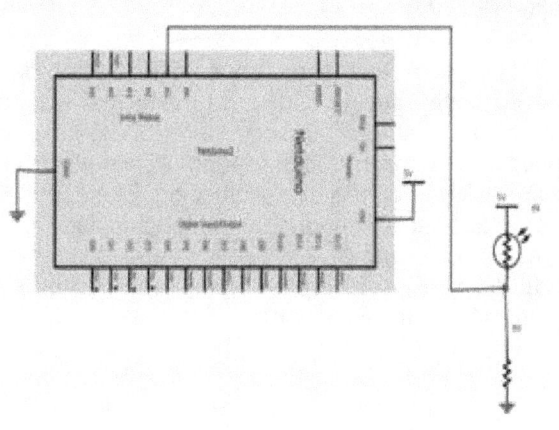

Netduino 2 en Español

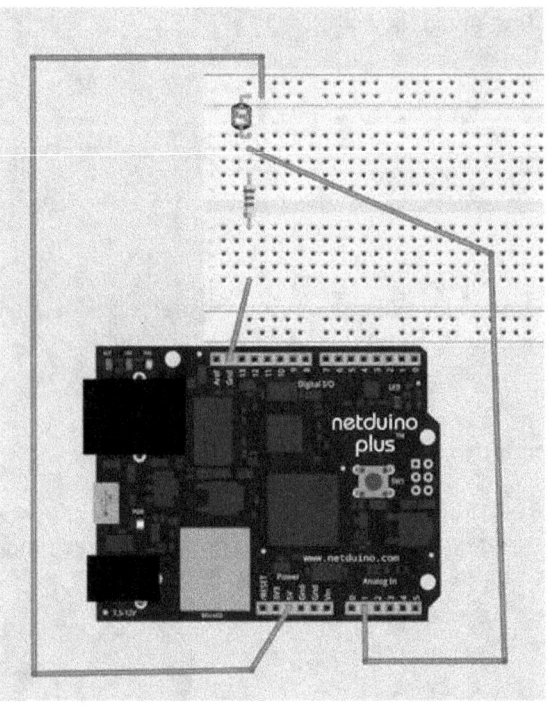

2-Ejemplo de aplicación con un LDR

En el siguiente código de ejemplo, capturaremos el nivel de luz, y si se sobrepasa un umbral máximo, apagaremos el led azul interno, así como sacaremos por la consola el mensaje "Se ha sobrepasado el umbral de luz".

En el ejemplo nuevamente definimos la variable voltagePortluz, como una instancia del objeto OutputPort, que hemos definido asociado al led interno, y con estado apagado por defecto al inicializarlo:

```
var ledPort = new OutputPort (Pins.ONBOARD_LED, false);
```

Como aspecto importante, definimos la variable **voltajePortluz** como una instancia del objeto AnalogInput, que hemos definido asociado al puerto 2 del conversor analógico- digital interno.

```
Var voltagePortluz = new Microsoft.SPOT.Hardware.AnalogInput
(Cpu.AnalogChannel.ANALOG_1); //sensor entrada analógica en A1 para sensor de luz
```

Teniendo definido el objeto, como se ha comentado en capítulos anteriores, para encender el led, basta utilizar el método **Write**

71

pasándole el valor **true** como único parámetro:

```
ledPort.Write (true);
```

Y de modo similar, para apagar el Led, bastará utilizar el método **Write**, pasándole el valor **false** como parámetro:

```
ledPort.Write (false);
```

La lectura de la señal externa lo realizarnos con el método **Read** sin parámetros, salvando el resultado en una variable del tipo double:

```
double rawValue;
...............
rawValue = voltagePortluz.Read ();
```

Respecto el método Debug.Print ("..."), como hemos visto en varios capítulos anteriores, lo empleamos en puntos del programa a afectos de depuración, para corroborar las acciones que se hagan sobre el led interno.

Por último, insertaremos todo el código dentro del cuerpo entre llaves *{}* en un bucle while (true), cuya condición no cambiará nunca, lo cual en la práctica permitirá que el led se encienda y se apague según la condición de ruido en un bucle infinito.

Para detectar cuando superamos un cierto umbral máximo de luz, utilizaremos una expresión condicional *if{ ..} else{..}*, donde lo destacable es que al estar dentro del bucle infinito, se está comparando constantemente el valor obtenido en el conversar A/D, con un valor de referencia que se ha estimado el 99.9% del tope. Si el lector lo desea puede experimentar con otros valores.

```
if (rawValue >= tope) // si sobrepasa un umbral
{
Debug.Print ("Se ha sobrepasado el umbral de luz");
ledPort.Write (false); // apaga led placa
}
else
{
ledPort.Write (true); // enciende led placa
}
```

Como peculiaridad, obsérvese en este ejemplo, que encendemos el led interno en forma inversa a como lo hemos hecho en otros ejemplos de capítulos anteriores, debido a que, como estamos midiendo un nivel de

Netduino 2 en Español

luz, no queremos que se perturbe la medida precisamente con la propia luz producida en la activación del citado led.

Para probar nuestro nuevo código, ya solo nos falta pulsar el botón **"Iniciar"** (botón que hay debajo del menú superior que pone COMPILAR), y acto seguido, tras la compilación, empezará a ejecutarse en Netduino 2, encendiendo o apagando el led interno en función del nivel de luminosidad detectado por el LDR, y también sacando un texto por la ventana de depuración de "Resultados".

A continuación mostramos el código completo de este ejemplo:

```csharp
using System;
using System.Net;
using System.Net.Sockets;
using System.Threading;
using Microsoft.SPOT;
using Microsoft.SPOT.Hardware;
using SecretLabs.NETMF.Hardware;
using SecretLabs.NETMF.Hardware.Netduino;

namespace Netduino Application10
{
    public class Sensorluz
    {
        public static void Main ()
        {
            //Sensor de luz By CRN @ 2014

            double tope = 0.99;
            double rawValue;
            string ttt;


            var ledPort = new OutputPort (Pins.ONBOARD_LED, false);

            var voltagePortluz = new Microsoft.SPOT.Hardware.AnalogInput
            (Cpu.AnalogChannel.ANALOG_1); //sensor entrada analógica en A1
            para sensor de luz


            Debug.Print ("Leyendo puerto...");

                while (true) //bucle infinito
                {
                rawValue = voltagePortluz.Read (); //captura valor sensor de
                humo
                ttt = rawValue.ToString ();
                Debug.Print ("valor =" + ttt);

                    if (rawValue >= tope) // si sobrepasa un umbral
                    {
                     Debug.Print ("Se ha sobrepasado el umbral de
                    luz");
```

```
                    ledPort.Write (false); // apaga led placa
                    }
                    else
                    {
                    ledPort.Write (true); // enciende led placa
                    }

        }


        }
    }
}
```

10 Sensor de movimiento

1-Introducción a los PIR

Todos los objetos con una temperatura por encima del cero absoluto emiten energía calorífica, lo cual significa que emiten radiación infrarroja, invisible para el ojo humano, pero que podemos detectar gracias a los sensores **PIR** (Passive Infrared Sensor), es decir, a los sensores pasivos de Infrarrojos.

El término pasivo se refiere que los **PIR** no generan ni irradian energía para fines de detección, basándose únicamente en la detección en la energía emitida por otros objetos: es decir los sensores PIR no detectan o miden el "calor" de por sí, sino que detectan la radiación infrarroja emitida por los objetos

Los **PIR** se construyen de modo que la radiación infrarroja entra a través de la parte frontal del sensor, conocida como la **cara del sensor.** El núcleo es de estado sólido y suele estar formado de compuestos piroeléctricos que generan energía cuando se exponen al calor, tomando la forma de una película delgada muy pequeña que suele ser parte de un circuito integrado.

Este C.I. suele ir alojado en una pequeña placa, que incluye el resto de electrónica, incluyendo generalmente una lente tipo fresnel de plástico que le proporciona un alcance de unos 5 metros y un ángulo de detección de 100º.

Los sensores PIR, son usados como detectores de movimientos en

domótica, instalaciones de seguridad y también en robótica gracias a su pequeño tamaño y bajo consumo.

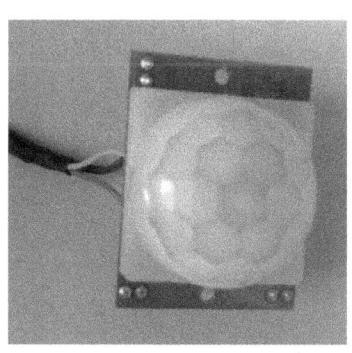

En este capítulo, usaremos como detector de movimiento el sensor HC-SR501, por ser uno de los sensores PIR más conocidos y usados en el mundo de la domótica, aunque la mayoría de los sensores discretos comerciales funcionan de un modo similar a éste.

El módulo HC-SR501 sólo mide 25 x 35 x 18 mm y, como la mayoría de sensores PIR, tiene 3 pines de conexión +VCC, OUT (3,3v) y GND, los cuales generalmente van marcados en la placa del circuito impreso.

Las características del módulo HC-SR501 son las siguientes:

- CI procesador de señal Sanyo genius BISS0001.
- Voltaje: 5V – 20V.
- Consumo: 65mA.
- Salida TTL : 3.3V, 0V.
- Tiempo de respuesta: ajustable (0.3 seg – 10 minutos).
- Tiempo de espera: 0.2 seg.
- Métodos de activación: L – disable repeat trigger, H enable repeat trigger.
- Radio de alcance: menos de 120 °, hasta 7 metros.
- Temperatura: – 15 ~ +70ºC.
- Dimensiones: 32*24 mm, distancia entre tornillos 28mm, M2, dimensiones de la lente en diámetro: 23mm.

Esta unidad funciona muy bien de 5 a 12 V, (su hoja de datos muestra 12V), aunque también puede funcionar con 3.3V consumiendo tan solo 1.6mA si se alimenta a esa tensión, llegando a consumir tan solo 350 uA mientras esta en reposo si se alimenta a 5V.

Para ajustar los tiempos de respuesta, cuenta el módulo en un lateral del circuito impreso, con dos resistencias variables de calibración (Ch1 y RL2) con las siguientes funciones:

- Ch1: Con esta resistencia podemos establecer el tiempo que se va a mantener activa la salida del sensor. Una de las principales limitaciones de este módulo es que el tiempo mínimo que se puede establecer es de unos 3 segundos. Si cambiamos la resistencia por otra de 100K, podemos bajar el tiempo mínimo a 0,5 segundos.

- RL2: Esta resistencia variable nos permite establecer la distancia de detección, la cual puede variar entre 3-7m.

La configuración de estos sensores, permite que múltiples sensores de movimiento puedan ser conectados en serie al mismo pin de entrada, pues si alguno de los sensores de movimiento se apaga, el pin de entrada será puesta a masa (0V) y podrá ser detectado por el circuito de control.
Por último, un aspecto importante de este módulo, es que la señal de salida o pin de salida es compatible TTL, y es en colector abierto, lo que significa que necesitaremos de una resistencia entre el positivo de la alimentación y la salida para leerla con Netduino 2+.

Una vez alimentado el PIR, debemos esperar entre 1-2 segundos para que el sensor pueda obtener una instantánea de la habitación, de modo que si algo se mueve después de ese período, el pin de la "alarma" pasará a estado BAJO y podrá ser detectado por Netduino 2+.

En el ejemplo que vamos a ver, alimentaremos el módulo con 5V, conectando además una resistencia de 10k entre la salida y la alimentación. La salida del módulo al ser discreta, a su vez lo conectaremos al pin 8 digital (marcado con D7), como podemos comprobar en el siguiente esquema eléctrico:

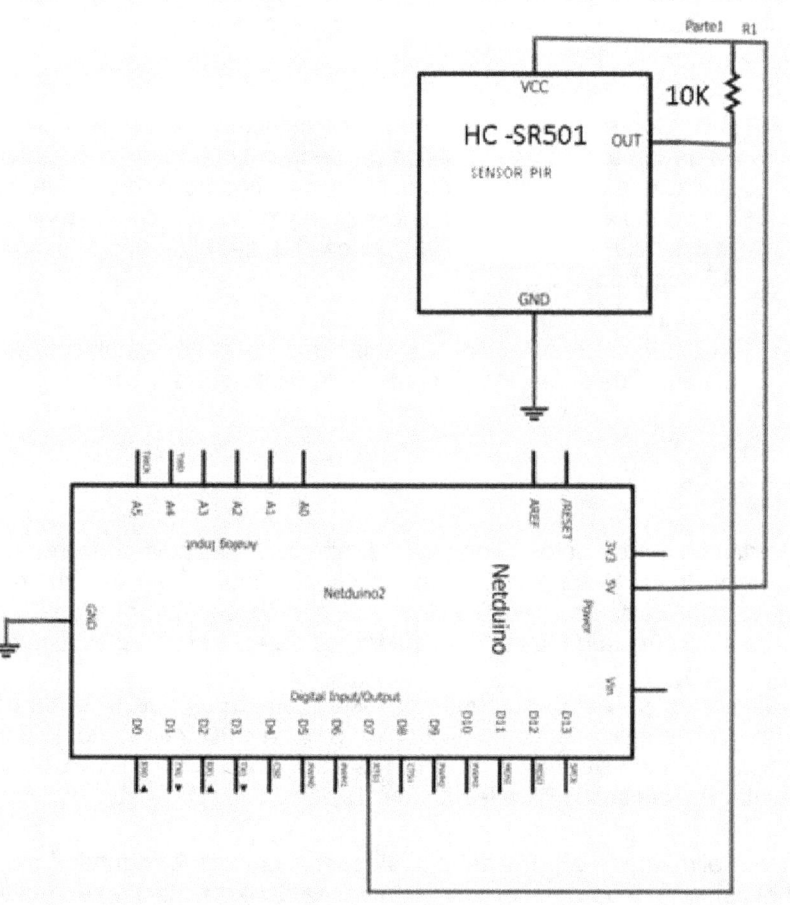

2-Ejemplo de aplicación con PIR

En el siguiente ejemplo, vamos a ver como se puede usar un sensor PIR con Netduino.

La lectura de señales binarias lo haremos a través del objeto InputPort, el cual deberá ser instanciado en una variable, en nuestro ejemplo, en la variable **pir**.

```
var pir = new InputPort (Pins.GPIO_PIN_D7, false, Port.ResistorMode.Disabled);
```

Una vez instanciado en una variable el objeto InputPort, la forma de

Netduino 2 en Español

recuperar el valor en ese momento de esa entrada binaria, es mediante el método **Read**, el cual se encargará de almacenar en una variable booleana el valor binario de esa dicho valor.

Como veremos en el ejemplo, la variable que almacenará el estado del sensor PIR es **estado** y para usarla, necesitaremos definirla previamente como booleana:

```
bool estado=false;
..........
estado = pir.Read ();
```

Asimismo, también definimos la variable **ledPort,** como una instancia del objeto OutputPort, que hemos definido asociado al led interno, y con estado apagado por defecto al inicializarlo:

```
var ledPort = new OutputPort (Pins.ONBOARD_LED, false);
```

Teniendo definido el objeto, como se ha comentado en capítulos anteriores, para encender el led, bastará utilizar el método *Write*, pasándole el valor **true** como único parámetro:

```
ledPort.Write (true);
```

Y de modo similar, para apagar el Led, bastará utilizar el método **Write** pasándole el valor **false** como parámetro:

```
ledPort.Write (false);
```

Para detectar cuando se produce movimiento, y por tanto la señal de alarma, utilizaremos un típico bloque condicional *if{..}else{..}*, donde lo destacable, es que al estar dentro del bucle infinito, se está constantemente leyendo el valor binario de la señal de estado de la señal de alama del sensor **PIR,** de modo que si detecta movimiento, activaremos el led interno.

Para probar nuestro nuevo código, ya solo nos falta pulsar el botón "Iniciar" (botón que hay debajo del menú superior que pone COMPILAR), y acto seguido, tras la compilación, empezará a ejecutarse el programa que hemos realizado en Netduino 2, encendiendo el led interno durante 1500ms y también sacando un texto por la ventana de depuración de "Resultados" en el caso de detectar objetos en movimiento.

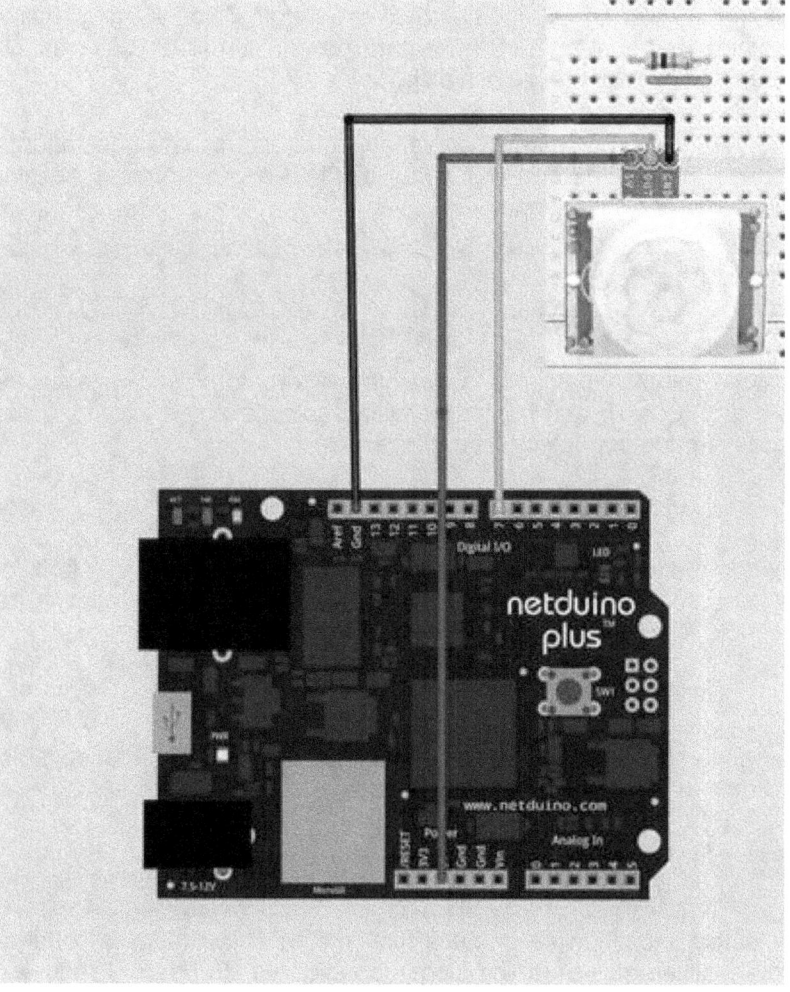

A continuación se muestra el código completo del ejemplo:

```
using System;
using System.Net;
using System.Net.Sockets;
using System.Threading;
using Microsoft.SPOT;
using Microsoft.SPOT.Hardware;
using SecretLabs.NETMF.Hardware;
using SecretLabs.NETMF.Hardware.Netduino;

namespace NetduinoApplication7
{
        public class Sensorsonido
```

```csharp
{
    public static void Main ()
    {
        //Sensor de obstáculos con PIR By CRN @ 2014
        bool estado = false;
        var ledPort = new OutputPort (Pins.ONBOARD_LED, false);
        var pir = new InputPort (Pins.GPIO_PIN_D7, false,
        Port.ResistorMode.Disabled);
        Debug.Print ("Leyendo puerto...");

        while (true) //bucle infinito
        {
            estado = pir.Read (); //captura valor sensor de humo

            if (estado) // si sobrepasa un umbral
            {
                Debug.Print ("Se ha detectado presencia");
                ledPort.Write (true); // enciende led placa
                Thread.Sleep (1500);
            }
            else
            {
                ledPort.Write (false); // apaga led placa
            }

        }

    }

}
```

11 Sensor de consumo eléctrico

1-Monitorización del consumo eléctrico

1.1-Medidas de seguridad

El circuito propuesto implica una alta corriente y voltaje, por lo que podría ser peligroso para la integridad del lector, así que por favor, siga las siguientes instrucciones:

- Mantenga siempre baja tensión (salida pulsos) y media tensión separados.
- Utilice cables con aislamiento y sección apropiados para la parte de media tensión.
- Utilice herramientas apropiados para el manejo de la media tensión.
- Asegúrese de desconectar de media tensión antes de manipular cualquier elemento.
- Si no está seguro de lo que está haciendo, por favor acuda o pregunte a personal cualificado.

1.2-Optimización del consumo eléctrico

El consumo energético supone un importante fuente de gasto para los ciudadanía siendo además, dado el origen de procedencia de la energía (básicamente de energías no limpias como la energía nuclear o de la combustión de combustibles fósiles), una fuente de contaminación cada vez mayor, pues al aumentar nuestra demanda energética, no solo aumentamos nuestra gasto energético, sino que también contribuimos

notablemente a emitir sustancias contaminantes a la atmósfera, contribuyendo así a aumentar el efecto invernadero.

Para intentar optimizar el consumo energético de cualquier entidad, es necesario antes medir éste de la forma más fiel posible, pues *no se puede optimizar nada que no se pueda medir*, siendo lo ideal monitorizar en tiempo real el consumo eléctrico de la instalación con objeto de: *crear alarmas, registrar eventos y crear informes de análisis de calidad de energía,* que son necesarios para proporcionar alertas al usuario, para impedir consumos excesivos y poder tener controlado el gasto energético.

Es importante, asimismo, destacar como estas políticas no solo repercuten en ayudarnos a reducir el costo del consumo eléctrico particular, sino también, nos permite como ciudadanos concienciados con el medio ambiente, ayudar a la sociedad a reducir la emisión de gases de efecto invernadero a la atmósfera.

La monitorización del consumo eléctrico nos posibilita, entre otras, las siguientes facilidades:

1. Estudiar los hábitos de consumo en función de las franjas horarias con objeto de promover políticas destinadas a su reducción.
2. Detectar picos indebidos de consumo en horas de baja actividad con objeto de detectar "consumos fantasmas" y consumos indebidos.
3. Racionalizar el consumo energético consumiendo solo lo necesario evitando derroches superfluos, contribuyendo de esta manera a luchar contra el cambio climático.
4. Predecir el consumo energético estudiando el histórico.
5. Trasladar los picos de consumo, cuando sea posible, a otras franjas horarias con objeto de no sobrecargar la red.
6. Detectar cortes de suministro con objeto de generar alarmas o tomar medidas correctivas.
7. Optimización de la demanda contratada.

Desglosando el consumo total eléctrico de una entidad, la iluminación podría llegar a representar hasta el 40% del consumo de energía en edificios, por lo que aplicar soluciones eficaces en su control, permitiría ahorrar fácilmente hasta el 50% de la factura de electricidad, en comparación con los métodos tradicionales.

Aun más importante en la cuota final, son los sistemas de climatización, los cuales pueden representar hasta el 70% del consumo de energía.

Según analistas, se podrían conseguir ahorros entre el 15 y el 30% en los costes energéticos, combinando algunos de los siguientes métodos:

- Programando el punto de ajuste de temperatura en función de la ocupación.
- Adaptando la potencia de calefacción o refrigeración, según las necesidades reales del edificio.
- Elevando la temperatura al nivel cómodo, cuando se detecte la presencia de ocupantes.
- Adaptando el flujo de ventilación, según la ocupación o el nivel de contaminación del aire interno.
- Recuperando energía de calefacción o refrigeración del aire extraído.

1.3-Medida del consumo eléctrico

Hasta hace muy poco, la medida de consumo se realizaba a través de resistencias o "**Shunts**", las cuales no dejan de ser resistencias de muy bajo valor y de alta disipación, las cuales se interponen en serie con el circuito a medir de modo que, por la ley de Ohm, la tensión en sus bornes es directamente proporcional a la intensidad global que atraviesa el circuito a medir. Como inconveniente, destacan que sus partes están expuestas (con el consiguiente peligro para las personas), necesitan interrumpir el circuito (y por tanto requiere cortar el suministro de c.a.) y además deben tener un consumo energético.

Modernamente se han introducido los **transformadores de corriente**, los cuales no son más que un transformador en el que se ha reemplazado un arrollamiento por unas pocas vueltas de un conductor de gran sección, el cual procede justamente del circuito a medir: por la ley de Lentz, la corriente inducida en el bobinado en el secundario, será proporcional a la corriente que atraviese por el citado conductor.

Para mejorar aún más el diseño y simplificar la instalación de éstos, se puede hacer el entrehierro del transformador movible (para no tener que interrumpir el circuito para hacerlo pasar por el transformador), como es el caso de las sondas amperimétricas.

Buscando una solución compacta, usando el principio anterior, existen vatímetros digitales de bajo coste, construidos en torno a un circuito serie y al que se ha añadido la electrónica necesaria (la salida del arrollamiento se conecta a un conversor D/A y a una lógica que nos muestre la lectura en w/h un pequeño display, ofreciendo además un salida de pulsos normalizada), introduciendo todo el conjunto en un carril

DIN, el cual se puede montar como un elemento más en el cuadro de distribución de corriente alterna, justo antes del circuito a medir (normalmente se conectará a la salida del elemento de corte y protección del circuito, y dimensionado a la potencia a medir).

En el caso de que estemos interesado en procesar con Netduino el consumo de una entidad (vivienda, local, circuito, etc.), se recomienda, por su bajo precio y altas prestaciones, unidas a su gran precisión, adquirir un vatímetro digital con pantalla LCD para instalar en carril que este dotado de salida óptica DIN 43864 (éste último es el estándar que define la interfaz asociado a la salida de pulsos).

En el ejemplo se ha usado el watimetro digital ADM255C, que como podemos ver, soporta 32 Amperios de carga máxima

Las características más importantes son las siguientes:

- Voltage operation range: 230V 32A max, 50/60Hz
- Accuracy class: 1
- Installation: DIN rail (DIN EN50022) ITE: 18mm (DIN 43880)
- Interface: optical coupler – DIN 43864
- Impuls length: 90ms(Ra: 0.5Wh/imp)
- Display: 6 + 1 = 999999.9 kWh With status LED
- Voltage operation range: 195-253 VAC; 0,025 ~ 30 A
- Starting current Cos-Phi=1: 20mA
- Frequency range: 50Hz or 60Hz +- 10%
- Temperature range: -20°C to +50°C
- Relative air humidity: to 75% – at short time value 95%

1.4-El estándar DIN 43864

La salida DIN 43864 define una salida opto aislada de la electrónica del vatímetro con objeto de no exponer al circuito que vayamos a usar a tensiones peligrosas que podrían dañarlo.

Dicha salida de pulsos, es en colector abierto, por lo que para leer dichos impulsos con Netduino 2+, conectaremos una resistencia de 10kohms entre masa y la salida SO- y un pin digital, conectando directamente el pin SO+ a VCC (pin 5V).

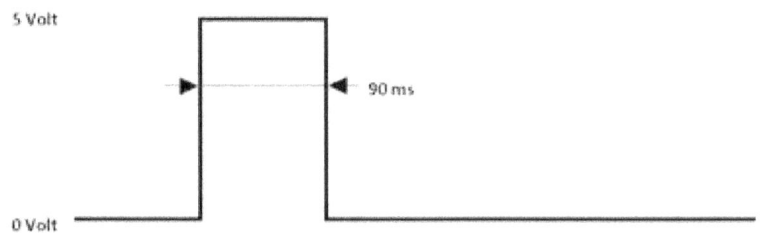

La longitud de los impulsos es de aproximadamente 90 ms, y estos son generados en este modelo cada 0.5Watios/hora, lo que significa que cada vez que contemos 1000 pulsos se habrá consumido 500Wat/hora (0.5KWH), y por tanto **para consumir un 1KWH, deberemos contar 2000 pulsos,** la cual será la fórmula que usaremos para implementar la lógica en el software que vamos a desarrollar.

2-Conexionado de un vatímetro digital

A la hora de realizar el conexionado de un vatímetro digital, debe tenerse muy especial cuidado el modo de realizarlo, dado que no todos lo vatímetros digitales se instalan del mismo modo, por lo que se recomienda que estudie primero la hoja de características y de montaje de su vatímetro antes de proceder a la instalación de éste. De hecho, en algunos nuevos modelos **se colocan en la parte inferior los dos neutros y en la parte superior se conectan los dos activos** (el de entrada de c.a. y la salida de utilización del circuito a medir), aunque en general, lo normal, es que respeten el esquema de conexión clásico de **conectar la entrada (fase y neutro) en la parte superior y la salida del circuito a medir en la parte inferior.**

En el caso de que tenga los conocimientos adecuados, y esté seguro de que no va a poner en peligro su integridad, el modo más habitual de insertar el vatímetro digital en el cuadro de distribución de corriente alterna, seria intercalando éste entre la salida del interruptor diferencial general y la entrada común de todos los magneto-térmicos asociados a los circuitos de utilización de la vivienda, tal y como se muestra en la siguiente figura:

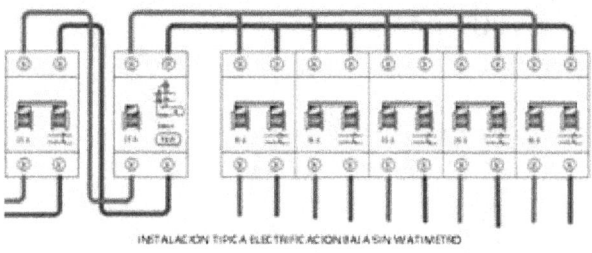

INSTALACION TIPICA ELECTRIFICACION BAJA SIN WATIMETRO

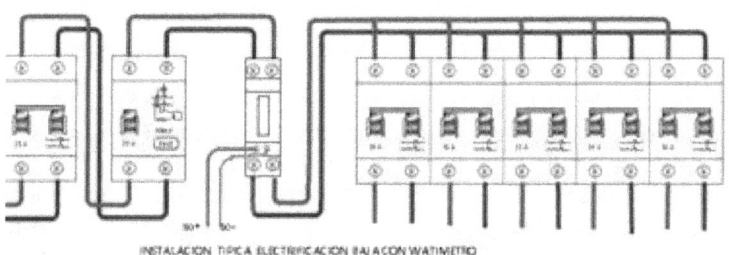

INSTALACION TIPICA ELECTRIFICACION BAJA CON WATIMETRO

Como se puede apreciar, del vatímetro digital, también conectaremos dos hilos de las respectivas dos salidas SO, la cual llevaremos a nuestro circuito para su procesamiento.

3-Ejemplo de aplicación

En el ejemplo que se propone, vamos a medir los pulsos procedentes de un vatímetro digital con salida de pulsos DIN 43864, concretamente el modelo DM255C comentado anteriormente, del cual llevaremos su salida de pulsos respectivamente marcada con S+ a la salida de +5v de la placa de Netduino 2+ y su salida S- al pin D0 la placa de Netduino 2+ ,conectando además una resistencia de 10 kilo-ohmios entre dicho terminal y la masa de Netduino 2 + (terminal marcado con 0V), siguiendo el esquema siguiente:

Netduino 2 en Español

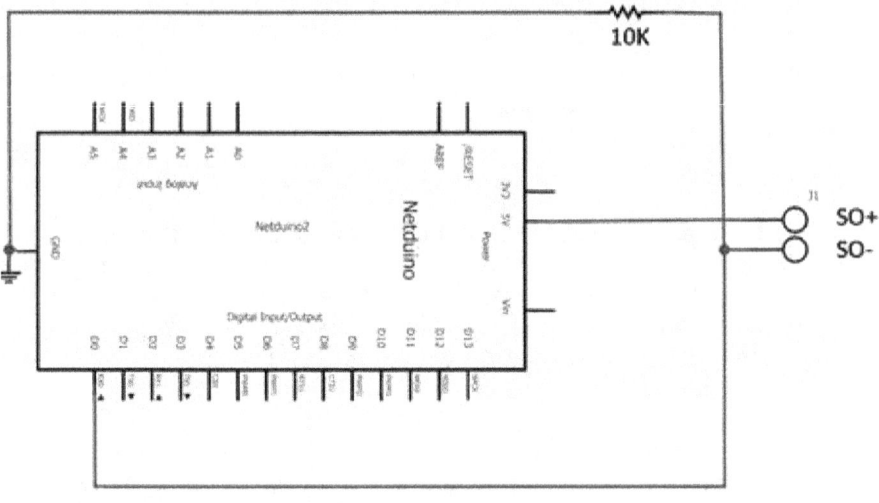

Una vez montado el circuito, la lectura de señales binarias lo haremos a través del objeto InputPort, el cual deberá ser instanciado en una variable, en nuestro ejemplo, en la variable "so".

```
var so = new InputPort (Pins.GPIO_PIN_D0, false, Port.ResistorMode.Disabled);
```

Una vez instanciado en una variable el objeto InputPort, la forma de recuperar el valor en ese momento de esa entrada binaria es mediante el método **Read,** el cual se encargará de almacenar en una variable booleana que deseemos el valor binario de esa dicho valor.

Como veremos en el ejemplo, esta variable es "**estado**" y se debe definir anteriormente como booleana.

```
bool estado=false;
..........
estado = so.Read ();
```

Asimismo, también definiremos, una vez, más la variable **ledPort,** como una instancia del objeto OutputPort, que hemos definido asociado al led interno ,y con estado apagado por defecto al inicializarlo.

```
var ledPort = new OutputPort (Pins.ONBOARD_LED, false);
```

Teniendo definido el objeto, como se ha comentado en capítulos anteriores, para encender el led, basta utilizar el método **Write,** pasándole el valor **true** como único parámetro:

```
ledPort.Write (true);
```

Y de modo similar, para apagar el led, bastará utilizar el método **Write** pasándole el valor false como parámetro:

```
ledPort.Write (false);
```

Para detectar cuando se ha producido un impulso del vatímetro, y por tanto se han consumido **0.5 Kw/h**, en esta primera versión del código (después veremos una forma mucha más eficiente ayudándonos de las interrupciones), utilizaremos una expresión condicional *if{ ...}else{..}*, en un bucle infinito, en el que se estará contantemente leyendo el valor binario de la señal de estado de la salida SO del vatímetro digital .

En caso de detectar un nivel alto, antes de sumar los 0.5 Kw/h al total, **nos aseguraremos de leer exactamente un único pulso** (recordemos al lector que la norma DIN nos asegura que el pulso tendrá una duración de 90ms), para lo cual, leeremos al menos dos lecturas entre un periodo de 90 ms.

Para implementar esta lógica, en primer lugar debemos leer el estado del puerto:

```
estado = so.Read (); //captura valor sensor de humo
```

También nos ayudaremos de la variable global *cuenta*, la cual almacenará el número de lecturas de niveles lógicos altos, con objeto de tomar lecturas correctas:

```
double cuenta = 0;
```

Una vez obtenida la lectura del puerto SO del vatímetro, en caso de ser un "uno binario", entraremos dentro de una expresión condicional *if{ ...}Else ...},* el cual estará constantemente reseteándose al llegar al valor 2, o en caso contrario, incrementándose, pues su misión es simplemente contar dos lecturas seguidas de nivel lógico alto.

Obsérvese que dentro del segundo condicional, en caso de ser el primer pulso, dado que el pulso del vatímetro es de 90ms, debemos hacer una pequeña pausa para evitar tomar lecturas incorrectas, para lo cual utilizamos el método **Sleep** ,perteneciente al objeto **Thread,** pasándole el valor de 90 (que es la duración del pulso):

```
Thread.Sleep (90);
```

Netduino 2 en Español

Con todas estas anotaciones, los dos condicionales se construirían con el siguiente código:

```
if (estado)
{
        if (cuenta > 2) // si sobrepasa un umbral
        {
        a = a + .5;
        Debug.Print ("Consumo acumulado " + a + "Kw/h");
        ledPort.Write (true); // enciende led placa
        Thread.Sleep (100);
        cuenta = 0;
        }
        else
        {
        cuenta++;
        Thread.Sleep (50);
        }
}
Else
{
    ..... // no es un uno binario
}
```

Solo queda comentar, que en caso de detectar el pulso (al contar dos unos binarios separados en 90ms), **sumaremos 0.5 Kw/h** al total, lo notificaremos por la consola, y además activaremos el led interno:

```
a=a+.5;
Debug.Print ("Consumo acumulado "+a + "Kw/h");
ledPort.Write (true); // enciende led placa
Thread.Sleep (100);
```

Para probar nuestro nuevo código, ya solo nos falta pulsar el botón "**Iniciar**" (botón que hay debajo del menú superior que pone COMPILAR), y acto seguido tras la compilación, empezará a ejecutarse en Netduino el programa descrito, encendiendo y apagando el led interno durante 100 ms. cada vez que se detecta un pulso procedente del vatímetro digital y también sacando un texto por la ventana de depuración de "Resultados".

A continuación, se adjunta el código completo del ejemplo:

```
using System;
using System.Net;
using System.Net.Sockets;
```

```csharp
using System.Threading;
using Microsoft.SPOT;
using Microsoft.SPOT.Hardware;
using SecretLabs.NETMF.Hardware;
using SecretLabs.NETMF.Hardware.Netduino;

namespace Netduino Application16
{
        public class Sensorwatt
        {
                public static void Main ()
                {
                //Sensor de pulsos By CRN @ 2014

                 double cuenta = 0;

                 bool estado = false;

                 var ledPort = new OutputPort (Pins.ONBOARD_LED, false);

                 var so = new InputPort (Pins.GPIO_PIN_D0, false,
                 Port.ResistorMode.Disabled);

                 double a = 0;

                 Debug.Print ("Leyendo puerto...");

                          while (true) //bucle infinito
                          {
                          estado = so.Read (); //captura valor sensor de consumo

                                  if (estado)
                                  {
                                          if (cuenta > 2) // si sobrepasa dos
                                          alternancias
                                          {
                                          a = a + .5;
                                          Debug.Print ("Consumo acumulado " +
                                          a + "Kw/h");
                                          ledPort.Write (true); // enciende led
                                          placa
                                          Thread.Sleep (100);
                                          cuenta = 0;
                                          }
                                          else
                                          {
                                          cuenta++;
                                          Thread.Sleep (50);
                                          }
                                  }
                                  else
                                  {
                                   ledPort.Write (false); // apaga led placa
                                  }

                          }
```

```
            }
        }
    }
}
```

4-Uso de Interrupciones con Netduino

Si bien el diseño anterior es completamente funcional, no es del todo adecuado eficiente para leer señales relativamente rápidas (pulsos), pues el algoritmo está comprobando constantemente el estado del puerto de entrada para llevar a cabo la suma del contador de Kw/h sobre la base de esa condición.

Un método más eficiente, es crear un controlador de eventos, gracias al uso de una interrupción asociada al puerto de lectura (PIN_D0), que hará que cuando ocurra un determinado evento que nosotros establezcamos a dicho puerto, se llame automáticamente a una rutina llamada "**rutina de interrupción**". Ésta rutina será la encargada de realizar las acciones necesarias que deseemos (en nuestro caso incrementar un contador de Kw/h consumidos).

Para definir un puerto de interrupción, instanciaremos el objeto **InterruptPort,** para lo cual se requieren cuatro argumentos:

- **Port-id:** Es el identificador para el puerto de interrupción, el cual en nuestro caso es Pins.GPIO_PIN_D0 (en el caso de que el puerto con el identificador especificado ya estuviese en uso, se produciría una excepción).
- **GlitchFilter:** si deseamos activar el filtro de Glitch, lo cual es la opción por defecto que nosotros usaremos.
- **ResistorMode:** Si se desea activar la resistencia asociada al puerto interno, en nuestro caso desactivado, al necesitar una resistencia externa para polarizar el colector abierto de la salida SO del vatímetro, por lo que lo configuraremos así Port.ResistorMode.Disabled.

Interrupt: define las condiciones necesarias para el puerto para generar una interrupción. Existen cinco modos de disparo en .net MicroFramework 4.2:

- Port.InterruptMode.InterruptEdgeHigh: el evento se disparará cuando el pin cambie de bajo a alto (no es funcional bajo Netduino 2 Plus).
- Port.InterruptMode.InterruptEdgeLevelHigh: el evento se disparará cuando el pin pase a nivel alto (no es funcional bajo Netduino 2 Plus).

- Port.InterruptMode.InterruptEdgeLevelLow: el evento se disparará cuando el pin pase a nivel bajo (no es funcional bajo Netduino 2 Plus).
- Port.InterruptMode.InterruptNone: el evento no se disparará.
- Port.InterruptMode.InterruptEdgeBoth: el evento se disparará cuando el pin cambie de bajo a alto o alto a bajo. Éste será el modo de funcionamiento que usaremos en el ejemplo, pues de este modo nos aseguramos de haber leído exactamente un pulso procedente del vatímetro.

En esencia, así definiremos el objeto pulso asociado al puerto D0 de Netduino:

```
InterruptPort pulso = new InterruptPort (Pins.GPIO_PIN_D0, true,
Port.ResistorMode.Disabled, Port.InterruptMode.InterruptEdgeHigh);
```

Una vez instanciado el objeto, tan sólo nos queda asociar al nuevo objeto (en nuestro caso **pulso**) a la rutina de interrupción (en el ejemplo la hemos llamado **pulso_OnInterrupt**):

```
pulso.OnInterrupt += new NativeEventHandler (pulso_OnInterrupt);
```

Definida la rutina de interrupción, es preciso que elaboremos el código que contendrá dicha rutina. A diferencia de las clases normales, una rutina de interrupción requiere obligatoriamente tres argumentos estrictos: **data1** del tipo uint, *data2* del tipo uint y **time** del tipo DateTime.

Respecto al cuerpo de la rutina, simplemente nos aseguraremos de leer exactamente un pulso, el cual lo componen dos interrupciones (el primero procedente del cambio de nivel bajo a nivel alto y el segundo de nivel bajo a nivel alto). Para hacer esto, nos aseguraremos de no incrementar la lectura de forma incorrecta, por lo que contaremos un pulso por cada dos interrupciones, ayudándonos de la variable global **cuenta**, la cual está definida justo después de la clase principal (**Program**).

```
private static double cuenta = 0;
```

Asimismo, dado que el contador de pulsos también debe ser accesible por la rutina de interrupción, es preciso definir la variable que almacenará los sucesivos pulsos del mismo modo.

```
private static double contadortotal = 0; //esta variable la incrementara la rutina de
interrupciones
```

Netduino 2 en Español

El comportamiento del cuerpo de la rutina de interrupción, por tanto, se puede resumir por tanto en un simple elemento condicional *if {} else {},* el cual estará constantemente reseteándose al llegar a 2, o en caso contrario, incrementándose.

```
if (cuenta > 2)
{ //solo al segundo pulso, debe incrementar la variable contadortotal
 cuenta = 0;
}
else
{//solo al segundo pulso
cuenta++;
}
```

El código completo de la rutina de interrupción sería el siguiente:

```
static void pulso_OnInterrupt (uint data1, uint data2, DateTime time)
{
        if (cuenta > 2) //solo al segundo pulso
        {
        contadortotal = contadortotal + 0.5; // va incrementando el total de
        Kw/h de forma acumulativa.
        Debug.Print ("Contadortotal=" + contadortotal);
        ledPort.Write (true); // enciende led placa
        Thread.Sleep (100);
        cuenta = 0;
        }
        else
        {
        cuenta++;
        }
}
```

Sólo nos queda hablar del cuerpo principal, en el cual impediremos abortar la ejecución del programa mediante el método **Sleep**:

```
Thread.Sleep (Timeout.Infinite);
```

Una alternativa al código anterior, es que en su lugar, el código realice una simple acción de forma indefinida, como por ejemplo apagar el led interno:

```
bool moreTime = true;
while (moreTime) //bucle infinito
{
ledPort.Write (false); // apaga led placa
Thread.Sleep (100);
}
```

Para probar nuestro nuevo código, ya solo nos falta pulsar el botón "**Iniciar**" (botón que hay debajo del menú superior que pone

Carlos Rodríguez Navarro

COMPILAR), y acto seguido, tras la compilación, empezará a ejecutarse éste en Netduino 2, encendiendo y apagando el led interno durante 50ms cada vez que se detecte un pulso procedente del vatímetro, y también sacando un texto por la ventana de depuración de "Resultados" con el valor del resultado del acumulado del contador de pulsos.

Este es el código completo de ejemplo de lectura de un vatímetro usando interrupciones:

```csharp
using System;
using System.Net;
using System.Net.Sockets;
using System.Threading;
using Microsoft.SPOT;
using Microsoft.SPOT.Hardware;
using SecretLabs.NETMF.Hardware;
using SecretLabs.NETMF.Hardware.Netduino ;

namespace Netduino Application17
{
        public class Program
        {
                private static double contadortotal = 0; //esta variable la incrementara
                la rutina de interrupciones
                private static double cuenta = 0;
                static OutputPort ledPort = new OutputPort (Pins.ONBOARD_LED,
                false);

                        public static void Main ()
                        {
                        //configuramos interrupciones para poder contar los pulsos
                        del vatímetro
                        InterruptPort pulso = new InterruptPort (Pins.GPIO_PIN_D0,
                        true,                              Port.ResistorMode.Disabled,
                        Port.InterruptMode.InterruptEdgeBoth);

                        pulso.OnInterrupt    +=    new    NativeEventHandler
                        (pulso_OnInterrupt);

                        Debug.Print ("*********************************");
                        Debug.Print ("** Empezando adquisición...*******");
                        Debug.Print ("*********************************");


                        bool moreTime = true;
                        while (moreTime) //bucle infinito
                        {
                        ledPort.Write (false); // apaga led placa
                        Thread.Sleep (100);
                        }

                        }

                        static void pulso_OnInterrupt (uint data1, uint data2,
                        DateTime time)
```

96

Netduino 2 en Español

```csharp
{
    if (cuenta > 2) //solo al segundo pulso
    {
    contadortotal = contadortotal + 0.0005; // va
    incrementando el total de Kw/h de forma
    acumulativa
    Debug.Print ("Total =" + contadortotal+ "Kw/h");
    ledPort.Write (true); // enciende led placa
    Thread.Sleep (100);
    cuenta = 0;
    }
    else
    {
    cuenta++;
    }
}
}
}
```

12 Manejo de un display LCD

1-Introducción al estándar HD44780

El famoso Controlador de LCD Hitachi HD44780 es uno de los controladores de visualización disponibles para matriz de puntos de visualización de cristal líquido más común.

Aunque Hitachi desarrolló el HD44780 específicamente para controlar pantallas LCD alfanuméricas con una interfaz sencilla para ser conectado a un microcontrolador o microprocesador de propósito general, muchos fabricantes de pantallas integradas empezaron a usar ese mismo controlador con sus productos, por lo que a día de hoy, éste es el estándar informal para este tipo de pantallas que se autodenominan "compatibles con HD44780".

Estas pantallas LCD se limitan a presentar texto ASCII, caracteres Kana japoneses, y algunos símbolos en dos líneas de 28 caracteres en blanco y negro utilizándose a menudo en maquinaria de todo tipo, por ejemplo electrodomésticos serie blanca, aparatos de consumo, etc. así como también en los diseños de miles de aficionados a la electrónica a lo largo de todo el mundo.

Las pantallas LCD que incluyen el chip HD44780, o en cualquier LCD equivalente con HD44780, vienen en un pequeño número de configuraciones estándar, siendo los tamaños más comunes 8x1 (una fila de ocho caracteres), 16 × 20 × 2, 2 × 20 y 4, aunque también se hacen tamaños personalizados más grandes con 32, 40 y 80 caracteres, y con 1, 2, 4 u 8 líneas, siendo el tamaño **más común el de 16 × 2**, que

es el que usaremos en nuestro ejemplo.

Un aspecto muy importante, es que los LCD de caracteres pueden venir **con sin luz de fondo e incluso de muy diferentes colores y tipos**, encareciendo normalmente el precio del display, pero mejorando mucho su visibilidad en ambientes oscuros.

La tensión de funcionamiento nominal para la retroiluminación led es normalmente de 5V en su brillo máximo, con oscurecimiento a voltajes más bajos, que dependen de detalles como el color del led, por lo que normalmente siempre se conectará una resistencia limitadora en serie con el positivo de alimentación (en el ejemplo se uso una resistencia de 220 ohmios 1/4W).

El estándar Hitachi también define una interfaz estándar de 16 contactos, normalmente con conexiones de borde de tarjeta en centros de 0,1 pulgadas / 2,54 mm, aunque los que no tienen luz de fondo puede tener sólo 14 pines, y pueden omitir los últimos dos pines para alimentar la retroiluminación.

La función de los pines suele ser la siguiente:

1. Tierra (0V).
2. VCC (3,3 a 5 V).
3. Ajuste de contraste (VO).
4. Register Select (RS). RS = 0: Comando, RS = 1: Datos.
5. Lectura / escritura (R / W). R / W = 0: Escribir, R / W = 1: Leer. Este pin es opcional debido al hecho de que la mayoría de las veces sólo querrán escribir en él y no leer. El uso general de este pin será pues usarlos permanentemente conectado a 0V (masa).
6. Reloj (Enable).
7. Bit 0: No se usa en la operación de 4 bits.
8. Bit 1: No se utiliza en la operación de 4 bits.
9. Bit 2: No se utiliza en la operación de 4 bits.
10. Bit 3: No se utiliza en la operación de 4 bits.
11. Bit 4.
12. Bit 5.
13. Bit 6.
14. Bit 7.
15. Luz de fondo del ánodo (+) si lo incluye.
16. Luz de fondo del cátodo (-) si lo incluye

La interfaz de HD44780 permite dos modos de funcionamiento: 8 bits y de 4 bits.

Usar el modo de 4 bits es más complejo, pero reduce el número de conexiones activas necesarias, y es el modo que usaremos en nuestro ejemplo. En este modo, el chip se inicia en modo de 8 bits, con el conjunto de instrucciones diseñadas para permitir el cambio sin necesidad de los cuatro pines de datos inferiores. Una vez en el modo de 4 bits, los datos de caracteres y el control se transfiere como pares de 4 bits en los pines de datos superiores, D4-D7.

2-Ejemplo de aplicación con un LCD

En este nuevo ejemplo, vamos a usar también un display LCD compatible con HD44780, de dos filas de 16 caracteres, que es el más extendido.

La asignación de pines del display a Netduino 2, será la siguiente:

Pin LCD	Función	Netduino
Pin 4	Selección de registro (RS)	D8
Pin 6	Reloj (Enable).	D9
Pin 11	Bit 4	D10
Pin 12	Bit 5	D11
Pin 13	Bit 6	D12
Pin 14	Bit 7	D13
Pin 3	Ajuste de contraste (VO)	gnd
Pin 5	Leer /Escribir (R/W).	gnd
Pin 1	ground	gnd
Pin16	ground	gnd
Pin15	Vcc(con resistencia 220 ohmios)	+5v
Pin 2	Vcc	+5V

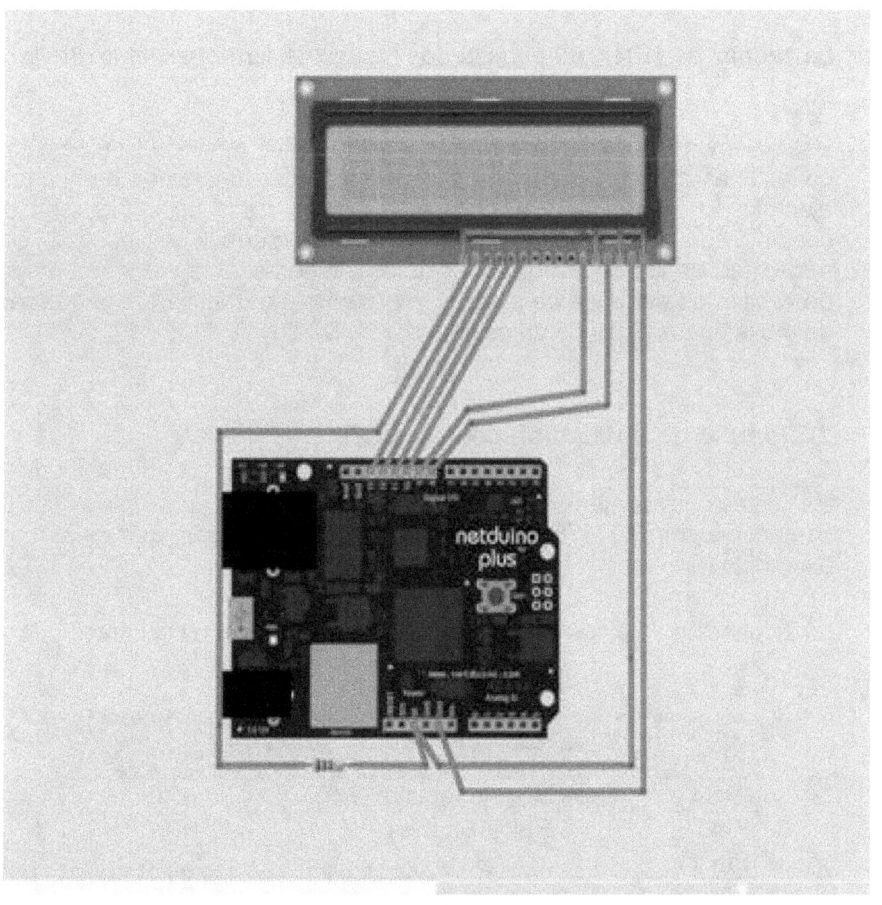

Una vez tengamos conectado el display a Netduino 2, cuyo montaje será similar a la imagen anterior, es momento de crear nuestra aplicación en c# que nos permita manejar dicho display, para lo cual, lo primero será obtener las librerías necesarias desde el site http://microliquidcrystal.codeplex.com/.

Una vez tengamos descargada las librerías, tenemos que salvarlas junto a nuestro proyecto.

Netduino 2 en Español

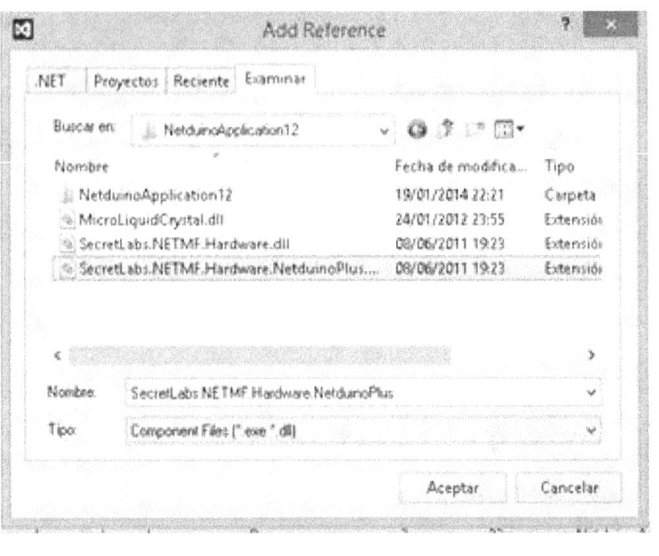

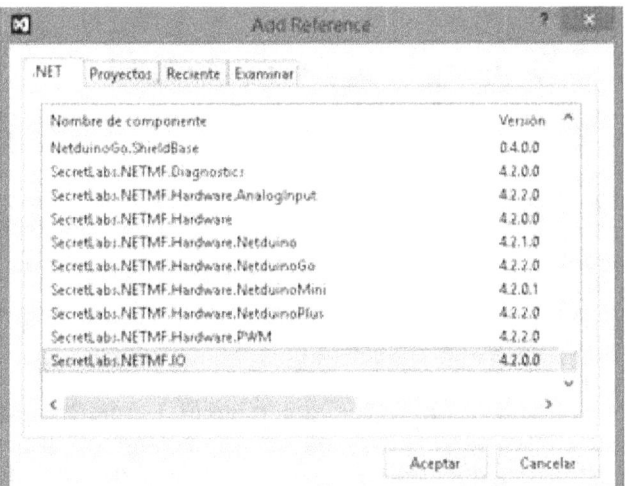

En el ejemplo que presentamos, además del LCD cuyas conexiones ya se han explicado, también hemos añadido al montaje una LDR, como se vio en el capítulo 9, conectado al pin analógico 1, tal y como se muestra en la siguiente imagen:

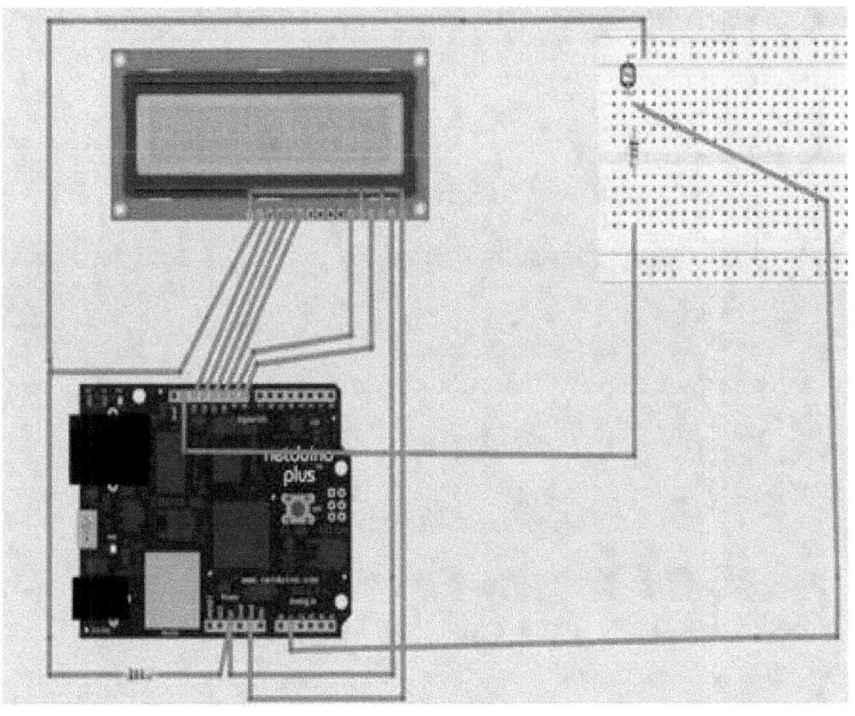

Para empezar a usar el LCD, hemos de hacer referencias a las librerías que hemos importado, las cuales nos van a permitir la conexión del LCD con nuestro Netduino.

```
using MicroLiquidCrystal;
using MicroLiquidCrystal.NETMF.Hardware.LCD;
```

Antes de crear el objeto LCD a través del método **LiquidCrystal,** debemos especificar cuáles serán los pines que conectaremos desde Netduino al LCD, cometido que realizaremos por medio del método GpioLiquidCrystalTransferProvider.

```
//configuración del LCD compatible con HD44780 de dos líneas
var lcdProvider = new GpioLiquidCrystalTransferProvider (
Pins.GPIO_PIN_D8, // RS
Pins.GPIO_PIN_D9, // enable
Pins.GPIO_PIN_D10, // d4
Pins.GPIO_PIN_D11, // d5
Pins.GPIO_PIN_D12, // d6
Pins.GPIO_PIN_D13); // d7
```

Una vez creada la instancia **lcdProvider** con los parámetros de nuestro LCD, ahora podemos llamar al constructor:

```
var lcd = new LiquidCrystal (lcdProvider);
```

Para hacer más sencillo el envío de caracteres a representar en el LCD, se ha creado un procedimiento **pintalcd**, al que pasaremos el objeto instanciado y el contenido de las dos filas de forma separada, como por ejemplo:

```
pintalcd (lcd, saludo2, l1);
```

El procedimiento, como vemos, requiere tres argumentos: el propio objeto **lcd,** el contenido de la **primera fila** a visualizar en el lcd (guardado en la variable **saludo2**) y el contenido de la **segunda fila** a visualizar en el lcd (contenido en la variable **l1**).

Para hacer agradable a la vista la representación de caracteres en el display LCD, vamos a usar una técnica de bajo refresco, por la que antes de enviar un mensaje a la primera o segunda fila, comprobaremos si este contenido ya está representado en el LCD, de modo que únicamente en caso de ser distinto para cada fila en particular, enviaremos la nueva cadena al LCD.

Esta técnica, se implementa simplemente almacenando en 2 variables para cada fila el valor antiguo y el nuevo, de modo que si son distintos se ejecute la acción:

```
if (l2 != l2old)
        {
        .....
        // envía un mensaje al LCD a la primera fila.
        }
l2old = l2;

if (l1 != l1old)
        {
        // envía un mensaje al LCD a la segunda fila. }

        }
l1old = l1;
```

En el código anterior, obsérvese que al final de cada **if** se iguala el valor antiguo con el valor nuevo, de modo que siempre que se llama a este procedimiento, actualizamos el valor antiguo al nuevo valor al salir de éste condicional.

Ahora veremos cómo nos comunicamos con el LCD.

Para "escribir" cualquier carácter en el LCD, primero debemos eliminar su contenido, acción que haremos invocando al método **Clear:**

```
lcd.Clear ();
```

Para indicar que el display es de 16 caracteres y dos filas, lo hacemos llamando al método **Begin**:

```
lcd.Begin (16, 2);
```

Ahora, en función de querer escribir en la primera fila o en la segunda fila, previamente debemos colocar el cursor en el sitio correspondiente.

Para escribir al principio de la primera fila escribiremos:

```
lcd.SetCursorPosition (0, 0);
```

Y para escribir en la segunda fila:

```
lcd.SetCursorPosition (0, 1);
```

Por último, para enviar caracteres al LCD, tanto para la primera o la segunda fila, simplemente invocaremos al método **Write** con la cadena a presentar en el display:

```
lcd.Write (l1);// envía un mensaje al LCD a la segunda fila. }
```

Para terminar este ejemplo, vamos a fijarnos en el cuerpo principal que está constituido básicamente por un bucle infinito **while (true),** el cual constantemente obtiene el valor de luz mediante **rawValue = voltagePortluz.Read ();** valor que compara con un valor umbral almacenado en la variable tope.

Si se sobrepasa ese umbral, se manda un mensaje al LCD de que se ha sobrepasado el umbral y se apaga el led interno, y en caso contrario, se envía simplemente el valor al LCD asegurándonos de dejar activo el led interno:

```
while (true) //bucle infinito que comprueba estado sensorhumo y puerta
{
          rawValue = voltagePortluz.Read (); //captura valor sensor de humo
          ttt = rawValue.ToString ();
          Debug.Print ("valor =" + ttt);

               if (rawValue == tope) // si sobrepasa un umbral
               {
               l2="Se ha sobrepasado el umbral de luz";
```

Netduino 2 en Español

```
pintalcd (lcd, l2, l1);//cada segundo actualiza el LCD
Thread.Sleep (1000);
ledPort.Write (false); // apaga led interno de la placa
}
else
{
 pintalcd (lcd, ttt, l1);//cada segundo actualiza el lcd
 Thread.Sleep (10);
ledPort.Write (true); // enciende el de led interno de la placa
}
```

```
} //del bucle infinito
```

Finalmente el código completo del ejemplo es el siguiente:

```
using System.Threading;
using Microsoft.SPOT.Hardware;
using SecretLabs.NETMF.Hardware.Netduino Plus;
using System.Text;
using System;

using System.Net;

using System.IO;

using Microsoft.SPOT;

using SecretLabs.NETMF.Hardware;

using System.Collections;
using System.Net.Sockets;
using MicroLiquidCrystal;
using Microsoft.SPOT.Net.NetworkInformation;
using Microsoft;
using System.Resources;
using MicroLiquidCrystal.NETMF.Hardware.LCD;

//using Toolbox.NETMF.NET;

public class SmartHome
{
        private static string saludo2 = "@";//
        private static string l1 = "Bienvenido";
        private static string l2old,l1old,l2 ;

        public static void Main ()
        {

        //configuración del LCD compatible con HD44780 de dos líneas
        var lcdProvider = new GpioLiquidCrystalTransferProvider (
```

```
Pins.GPIO_PIN_D8, // RS
Pins.GPIO_PIN_D9, // enable
Pins.GPIO_PIN_D10, // d4
Pins.GPIO_PIN_D11, // d5
Pins.GPIO_PIN_D12, // d6
Pins.GPIO_PIN_D13); // d7
var lcd = new LiquidCrystal (lcdProvider);
double tope = 0.0;
double rawValue;
string ttt;
var ledPort = new OutputPort (Pins.ONBOARD_LED, false);

var voltagePortluz = new Microsoft.SPOT.Hardware.AnalogInput
(Cpu.AnalogChannel.ANALOG_1); //sensor entrada analógica en A1 para
sensor de luz
Debug.Print ("Leyendo puerto...");
pintalcd (lcd, saludo2, l1);//cada segundo actualiza el lcd

            while (true) //bucle infinito que comprueba estado sensorhumo y
            estado puerta
            {
            rawValue = voltagePortluz.Read (); //captura valor sensor de humo
            ttt = rawValue.ToString ();
            Debug.Print ("valor =" + ttt);

                    if (rawValue == tope) // si sobrepasa un umbral
                    {
                    l2="Se ha sobrepasado el umbral de luz";
                    pintalcd (lcd, l2, l1);//cada segundo actualiza el lcd
                    Thread.Sleep (1000);
                    Debug.Print ("Se ha sobrepasado el umbral de luz");
                    ledPort.Write (false); // apaga led interno de la placa
                    }
                    else
                    {
                      pintalcd (lcd, ttt, l1);//cada segundo actualiza el lcd
                    Thread.Sleep (10);
                    ledPort.Write (true); // enciende el de led interno de la placa
                    }
            } //del bucle infinito
}

    static void pintalcd (LiquidCrystal lcd, string l1, string l2)
    {
            if (l2 != l2old)
            {
            lcd.Clear ();
            lcd.Begin (16, 2);
            lcd.SetCursorPosition (0, 0);
            lcd.Write (l2);// envía un mensaje al LCD a la primera fila.
            }
            l2old = l2;
            //Debug.Print (l2);

            if (l1 != l1old)
            {
            lcd.SetCursorPosition (0, 1);
            lcd.Write (l1);// envía un mensaje al LCD a la segunda fila. }
```

    i1old = i1;
    // Debug.Print (i2);
}
}//FIN

13 Servidor avanzado

1-Servidor NeonMika.Webserver

Gracias al austriaco Markus VV, que ha puesto a disposición de forma altruista su servidor web **NeonMika.Webserver @** para Netduino 2+ en http://neonmikawebserver.codeplex.com/, todos podemos disfrutar de un excelente servidor simplemente anexando su código, el cual ha puesto a disposición de forma gratuita en **codeplex.com** para poderlo usar en nuestros propios proyectos.

NeonMika.Webserver, que es el nombre del proyecto, es un servidor web pre-configurado muy fácil de extender que necesita una mínima (o ninguna) línea de código para poder lograr grandes resultados, permitiendo controlar su Netduino 2+, acceder y subir archivos, crear web-services y mucho más.

Entre las posibilidades de este excelente servidor, destaca el soporte de acceso a la tarjeta SD, control del Netduino 2+ utilizando métodos existentes como setPWM, setDigitalPinState, etc. e incluso permite usar nuestros propios métodos de servicio web.

Para probar NeonMika.Webserver, sólo tiene que seguir los siguientes pasos:

1. Descargar el código de aquí:
    http://neonmikawebserver.codeplex.com/downloads/ get/785664 (necesitará registrarse previamente en codeplex).

2. Descomprimir el fichero en un directorio de su ordenador.
3. Ejecutar Visual Studio.
4. Cargar el fichero de solución que está en la ruta de "**Executables**".(..\neonmikawebserverV1.2\Executeable\ NeonMikaWebserver\NeonMilkaWebServer.slh)
5. Personalizar **NeonMikaWebServerExecutable. Program.cs** ya que sólo tiene que llamar al constructor para iniciar NeonMika.Webserver.
Los parámetros que se necesitan para llamar al servidor son:

- En que puerto conectaremos el led de actividad.
- El puerto de su Netduino para el servidor web.
- DHCP activado / desactivado (true/false).
- Dirección IP de su Netduino.
- Máscara de subred.
- Dirección IP de la pasarela o Gateway.
- Nombre en red de su Netduino.

Por ejemplo para una dirección IP fija de Netduino 192.168.1.35, este sería el código:

```
using System.Threading;
using NeonMika.Webserver;
using NeonMika.Webserver.Responses;
namespace NeonMikaWebserverExecuteable
{
public class Program
{
public static void Main ()
{
Server WebServer = new Server
(PinManagement.OnboardLED,80,false,"192.168.1.35","255.
255.255.0","192.168.178.1","NETDUINOPLUS");
WebServer.AddResponse (new XMLResponse ("wave", new
XMLResponseMethod (WebserverXMLMethods.Wave)));
}
}
}
```

6. Compilar la solución (F7).
7. Observar la consola de Visual Studio, pues además de ver la actividad del servidor también veremos al final del log la dirección IP de nuestro servidor web (**ver anexo A**).

2-Configuración de red de Netduino 2 Plus

Para averiguar la dirección IP con la que nos comunicaremos con el servidor web, bien podemos forzar a Netduino 2+ a que tome una determinada dirección IP, o bien, podemos investigar cual es la dirección IP servida por el servidor DHCP.

En ambos casos, para configurar la dirección IP de Netduino 2+, deberemos hacerlo usando una herramienta incluida en todas las versiones de Microsoft.NET Framework llamada **MFDeploy.** Esta herramienta suele estar incluida dentro del directorio del Framework en Tools (por ejemplo en v4.3 estará en "C:\Program Files (x86)\Microsoft .NET Micro Framework\v4.3\Tools\MFDeploy.exe").

Una vez lanzada la herramienta Mdfdeploy, nos iremos a **Device**, seleccionado **USB,** y en ese momento, si tenemos conectado Netduino 2+ a nuestro ordenador, nos debería aparecer en el combo de la derecha un ítem correspondiente a la versión de nuestro Netduino.

Para averiguar la dirección Ip que está utilizando Netduino 2+, simplemente nos iremos al menú **Target** y pincharemos en **Device capabilities**, y en seguida nos mostrará en la primera línea la dirección Ip de Netduino 2+.

3-Métodos soportados en NeonMika Web server

Obtenida la dirección IP de Netduino 2+, asegurándonos que el servidor web se está ejecutando correctamente y tiene conexión a la red (led azul

interno parpadeando en ráfagas lentas), suponiendo que la dirección Ip de Netduino 2+ sea 192.168.1.35, vamos a mostrar una lista con todos los métodos web pre-codificados de NeonMika Web Server que se pueden utilizar, tanto dentro de su navegador, como desde cualquier otra aplicación, para comunicarse con su Netduino 2+:

GESTIÓN DE SEÑALES DIGITALES

switchDigitalPin

Funcionalidad:
Cambia el estado lógico del pin seleccionado de verdadero a falso o viceversa.
Sintaxis:
http://ip_Netduino/ switchDigitalPin pin = [0-13]
Ejemplo:
Petición1ª:http://192.168.1.35/switchDigitalPin?pin=0
Respuesta Netduino 2+:

This XML file does not appear to have any style information associated with it. The document tree is shown below.
```
<Response>
<pin0>1</pin0>
</Response>
```

Petición2ª:http://192.168.1.35/switchDigitalPin?pin=0
Respuesta Netduino:

This XML file does not appear to have any style information associated with it. The document tree is shown below.
```
<Response>
<pin0>0</pin0>
</Response>
```

setDigitalPin

Funcionalidad:
Asigna al valor del pin digital [0-13] seleccionado al estado que se le pasa como parámetro [true | false].
Sintaxis:
http://ip_Netduino/ setDigitalPin?pin = [0-13] y state = [true | false]
Ejemplo:
Petición: http://192.168.1.35/setDigitalPin?pin=0&state=true
Respuesta Netduino 2+:

This XML file does not appear to have any style information associated with it. The document tree is shown below.
```
<Response>
<pin0>1</pin0>
</Response>
```

getDigitalPinState

Funcionalidad:
Devuelve el estado de su pin seleccionado (on / off)).
Sintaxis:
http://ip_Netduino/ getDigitalPinState?pin = [0-13]
Ejemplo:
Petición:http://192.168.1.35/getDigitalPinState?pin=0
Respuesta Netduino 2+:

This XML file does not appear to have any style information associated with it. The document tree is shown below.
```
<Response>
<pin0>1</pin0>
</Response>
```

getDigitalPinState

Funcionalidad:
Devuelve el estado de cada pin digital.
Sintaxis:
http://ip_Netduino/getAllDigitalPinStates
Ejemplo:
Petición: http://192.168.1.35/getAllDigitalPinStates
Respuesta Netduino 2+:

This XML file does not appear to have any style information associated with it. The document tree is shown below.
```
<Response>
<pin11>0</pin11>
<pin2>1</pin2>
<pin10>0</pin10>
<pin9>0</pin9>
<pin1>0</pin1>
<pin13>0</pin13>
<pin4>0</pin4>
<pin7>0</pin7>
<pin6>0</pin6>
<pin5>0</pin5>
<pin8>0</pin8>
<pin0>1</pin0>
<pin12>0</pin12>
<pin3>0</pin3>
</Response>
```

GESTIÓN DE SEÑALES ANALÓGICAS

getAnalogPinValue

Funcionalidad:
Devuelve el valor del pin analógico seleccionado.
Sintaxis:
http://ip_Netduino/ getAnalogPinValue?pin = [0-5]
Ejemplo:
http://192.168.1.35/getAnalogPinValue?pin=0

Respuesta Netduino 2+:

```
This XML file does not appear to have any style information associated with it. The document tree is shown below.
<Response>
<pin0>185</pin0>
</Response>
```

getAllAnalogPinValues

Funcionalidad:
Devuelve el valor de cada pin analógico.
Sintaxis:
http://ip_Netduino/ getAllAnalogPinValues
Ejemplo:
Petición:http://192.168.1.35/getAllAnalogPinValues
Respuesta Netduino 2+:

```
This XML file does not appear to have any style information associated with it. The document tree is shown below.
<Response>
<pin2>145</pin2>
<pin5>182</pin5>
<pin1>123</pin1>
<pin4>160</pin4>
<pin0>92</pin0>
<pin3>165</pin3>
</Response>
```

Pwm

Funcionalidad:
Ajusta el generador por modulación de ancho de pulsos (PWM) para el pin especificado (recuerde que sólo se admiten los puertos 5,6,9 o 10) ,así como el período y la duración definidos.
Sintaxis:
http://ip_Netduino/ PWM pin = [5 | 6 | 9 | 10] y periodo = [int] y duración = [int]
Ejemplo:
Petición:http://192.168.1.35/pwm?pin=5&period=55&duration=10
Respuesta Netduino 2+:

```
This XML file does not appear to have any style information associated with it. The document tree is shown below.
<Response>
<success>55/10</success>
</Response>
```

getAllPWMValues

Funcionalidad:
Devuelve los valores para todos los puertos PWM.
Sintaxis:
http://ip_Netduino/ getAllPWMValues

Netduino 2 en Español

Ejemplo:
Petición:http://192.168.1.35/getAllPWMValues
Respuesta Netduino 2+:

This XML file does not appear to have any style information associated with it. The document tree is shown below.
```
<Response>
<pin5_period>0</pin5_period>
<pin6_period>0</pin6_period>
<pin10_period>0</pin10_period>
<pin9_duration>0</pin9_duration>
<pin10_duration>0</pin10_duration>
<pin5_duration>0</pin5_duration>
<pin6_duration>0</pin6_duration>
<pin9_period>0</pin9_period>
</Response>
```

UTILIDADES

echo

Funcionalidad:
Devuelve el valor enviado pasado como parámetro.
Sintaxis:
http://ip_Netduino/echo?value=[cadena]
Ejemplo:
Petición:http://192.168.1.35/echo?value=hola
Respuesta Netduino 2+:

This XML file does not appear to have any style information associated with it. The document tree is shown below.
```
<Response>
<echo>hola</echo>
</Response>
```

/SD/

Funcionalidad:
Permite inspeccionar el contenido de una microSD que se haya insertado en el slot de Netduino, presentándose los archivos y directorios como respuesta, ofreciendo además la posibilidad de descargar éstos, así de como navegar por el árbol de directorios (si los hubiese).
Sintaxis:
http://ip_Netduino/SD/(*)
(*)Sólo tiene que escribir la IP de su Netduino 2+, la barra "/" y el directorio o fichero y usted podría ver o descargar el archivo o directorio que desee. Si la ruta dada es un directorio, se devolverá una vista de directorio

Ejemplo:
Petición:http://192.168.1.35 /SD/

Respuesta Netduino 2+:

```
One level up
SD/
Directories:
\SD\System Volume Information
Files.
\SD\PRUEBA.HTM
\SD\OFF.HTM
\SD\ON.HTM
\SD\PRUEBA1.HTM
```

xmlResponselist

Funcionalidad: Da una lista de todos los métodos XML.
Sintaxis: http://ip_Netduino/xmlResponselist
Ejemplo:
http://192.168.1.35 / xmlResponselist

jsonResponselist

Funcionalidad:
Da una lista de todos los métodos JSON implementados en NeonMika Web-Server.
Sintaxis:
http://ip_Netduino/jsonResponselist
Ejemplo:
http://192.168.1.35/jsonResponselist

multipleXML

Funcionalidad:
En NeonMika Web-Server, el autor ha implementado un ejemplo de cómo usar XML anidados. Para invocarlo solo hay que invocar vía http a dicho método
Sintaxis:
http://ip_Netduino/multipleXML
Ejemplo:
Petición:http://192.168.1.35/multixml
Respuesta Netduino 2+:

```
This XML file does not appear to have any style information associated with it. The document tree is shown
below.
<Response>
<Phones>
<Phones ExampleAttribute2="1992" ExampleAttribute1="NeonMika">
<Phone>
<Name>Mokia Rumia</Name>
<PhoneNumber>436603541897</PhoneNumber>
<WirelessConnections>
<WLAN>False</WLAN>
<Bluetooth>True</Bluetooth>
</WirelessConnections>
```

Netduino 2 en Español

```
</Phone>
<Phone>
<Name>Langsum Talaxy</Name>
<PhoneNumber>436603541122</PhoneNumber>
<WirelessConnections>
<WLAN>False</WLAN>
<Bluetooth>True</Bluetooth>
</WirelessConnections>
</Phone>
</Phones>
</Phones>
<UseTheHashtable>If you don't need nested XML</UseTheHashtable>
</Response>
```

[]

Funcionalidad:
Ofrecer ayuda sobre la sintaxis de todos los métodos disponibles en NeonMika Web-Server.

Sintaxis:
http://ip_Netduino

Ejemplo:
Petición

http://192.168.1.35

Respuesta Netduino 2+:

NeonMika.Webserver

Hello out there!
Thanks for using NeonMika.Webserver! :)
The more people using this webserver, the more we can make it better.
So please give feedback at
http://forums.Netduino.com/index.php?/topic/2889-neonmikawebserver/
or/and at
http://neonmikawebserver.codeplex.com/
Thanks & have fun with NeonMika.Weberserver,
Markus

Here is a list with all pre-coded webmethods you can use
within your browser or any other application
to communicate with your Netduino:
echo (Returns the submitted value)
-> Netduinoplus/echo?value=[a-Z]
switchDigitalPin (Switches the selected pin from true to false and vis-a-vis)
-> Netduinoplus/switchDigitalPin?pin=[0-13]
setDigitalPin (Set the selected digital pin to the selected state)
-> Netduinoplus/setDigitalPin?pin=[0-13]&state=[true|false]
pwm (Set the PWM of the pin to the submitted period & duration
-> Netduinoplus/pwm?pin=[5|6|9|10]&period=[int]&duration=[int]
getAnalogPinValue (Return the value of the selected analog pin)
-> Netduinoplus/getAnalogPinValue?pin=[0-5]

getDigitalPinState (Returns your selected pin's state (on / off))
-> Netduinoplus/getDigitalPinState?pin=[0-13]
getAllAnalogPinValues (Return the value for each analog pin)
-> Netduinoplus/getAllAnalogPinValues
getDigitalPinState (Returns the state for each digital pin)
-> Netduinoplus/getAllDigitalPinStates
fileUpload (Uploads a file to the path on the SD card via POST. You have to write the file-data (bytes) into the POST
body)
-> Netduinoplus/upload?path=[a-Z]
AND FOR SURE:
File and directory response
Just type in Netduinoplus/[pathtomyfile] and you can view / download your file. If the given path is a directory, a directory
view will be returned
More for testing purpose, but also part of NeonMika.Webserver:
xmlResponselist (Gives you a list of all XML methods)
-> Netduinoplus/xmlResponselist
```

119
```

Carlos Rodríguez Navarro

jsonResponselist (Gives you a list of all JSON methods)
-> Netduinoplus/jsonResponselist
multipleXML (Example on how to use nested XML)
-> Netduinoplus/multixml

14 MIT App Inventor

1 Instalación de MIT App Inventor

Mit App inventor es un framework creado inicialmente por el MIT (Instituto tecnológico de Massachusetts), siendo adoptado después por Google, para que cualquier persona con interés, pueda crearse su propia aplicación móvil basada en Android, ya sea para su empresa, para su casa o por otros intereses.

Mit App Inventor es, por tanto, una herramienta basada en la nube creada por el MIT para la creación rápida de aplicaciones móviles en Android, lo que significa que usted puede construir aplicaciones directamente en su navegador web para su dispositivo Android, sin tener que escribir una sola línea de código en Java, y con el potencial de poder interactuar con Netduino 2+, como vamos a ver en el último capítulo de éste libro.

El propio sitio web de MIT ofrece todo el apoyo que necesitará a medida que aprenda a cómo construir sus propias aplicaciones, pero en este capítulo se intentarán ofrecer las bases para que usted mismo pueda modificar y mejorar la aplicación que vamos a mostrar que interactúa con Netduino 2+, o incluso crear otras nuevas aplicaciones que puedan comunicarse con Netduino 2+.

Puede configurar la aplicación MIT App Inventor y empezar a crear aplicaciones en cuestión de minutos, gracias a que todo el software está disponible en la nube o "servicio", en http://aai2.appinventor.mit.edu, y por tanto, no tiene que instalar nada en su ordenador, dado que el

Diseñador y Editor de bloques se ejecutan por completo en un navegador, abstrayendo de esta forma la complejidad al usuario (es por eso que también se dice que se ejecuta en "la nube").

Para probar su aplicación en un dispositivo mientras lo construye (también llamada "Testing en vivo"), en App Inventor puede seguir algunos de los tres siguientes métodos:

1- Construir aplicaciones con un dispositivo Android y conexión WiFi sin instalación de ningún software en su equipo. Es la opción recomendada, si está utilizando un dispositivo Android, y tiene una conexión wifi a Internet entre ambos, por la que usted puede comenzar la creación de aplicaciones sin necesidad de descargar ningún software en su ordenador.

Para empezar a programar con App inventor, solo necesitará instalar la aplicación **Companion App Inventor** en su Smartphone o tableta, o bien usar cualquier **lector de códigos QR,** por ejemplo el lector BIDI (es decir, cualquier programa que sirva para capturar el código QR generado por el software de APP Inventor).

El código QR aparecerá directamente en su pantalla cuando haya creado su aplicación en App Inventor, y haya ido a la opción **Build,** seleccionado **"App (provide QR code for .apk)".**

En caso de tener instalado **Companion App Inventor** en su Smartphone, la instalación de su app en su smartphone se limitará a escanear el código QR y aceptar la instalación del apk.

Esta es la opción que recomienda encarecidamente el MIT.

2-Usar un emulador: Si usted no tiene un dispositivo Android, usted tendrá que instalar un software en su ordenador para que pueda utilizar el emulador de Android en la pantalla. Por ejemplo, si es usted profesor y tiene una clase de 30 alumnos, puede hacer que éstos trabajen principalmente en emuladores y compartan unos pocos dispositivos.

3-Construir aplicaciones con un dispositivo Android y Cable USB. Si usted no tiene una conexión WIFI a Internet,pero si tiene un terminal Android, usted puede instalar el software en su ordenador de modo que usted pueda conectarse a su dispositivo Android a través de USB.

Netduino 2 en Español

La configuración de una conexión USB puede ser difícil, especialmente en máquinas Windows, que necesitan un controlador especial para conectarse a los dispositivos Android.

Desafortunadamente, diferentes dispositivos pueden requerir diferentes drivers, y puede que tenga que buscar en la web para encontrar el controlador apropiado para su Smartphone.

App Inventor ofrece un programa de prueba que verifica si el dispositivo conectado por USB puede comunicarse con el ordenador. Debe ejecutar esta prueba y resolver los problemas de conexión antes de intentar utilizar App Inventor con USB en ese dispositivo.

Por tanto, para conectar con USB, es necesario instalar primero el software de instalación de App Inventor en el equipo.

NOTA: App Inventor 2 no funciona con Microsoft ® Internet Explorer por el momento. Para los usuarios de Windows, el MIT recomienda utilizar **Google® Chrome,** o <u>Firefox,</u> como navegadores para utilizar con App Inventor.

A continuación, se explica cómo realizar los ajustes correspondientes, tanto en su Smartphone como en su ordenador, a la hora de desarrollar con Mit App Inventor:

1.1-Conexión a través de WiFi

Usted puede utilizar App Inventor sin descargar nada en su ordenador, como se comentaba al principio de este capítulo, si tanto su terminal Android como su ordenador, estan conectados a la misma red WIFI, por lo que para probar, se recomienda desconectar la red móvil en su Smartphone o Tableta y solo usar la conectividad wifi.

Los siguientes pasos le guiarán a través del proceso:

Paso 1: Descargar e instalar el MIT EA2 App Companion en su Smartphone o tableta.
Si tiene acceso a Google Play busque "**MIT AI2 Companion**" o vaya a la siguiente url en su terminal Android:
https://play.google.com/store/apps/details?id=edu.mit.appinventor.aicompanion3.
Si no tiene acceso a Google Play todavía puede instalar directamente la aplicación desde la siguiente url:
http://appinventor.mit.edu/ai2/MITAI2Companion.apk.

Para poder instalar este último, tendrá que habilitar una opción en la configuración de su dispositivo "**para permitir la instalación de aplicaciones desde fuentes desconocidas**"(*)(**).

(*)Para encontrar esta configuración en las versiones de Android anteriores a la 4.0, vaya a "Ajustes▶ Aplicaciones" y luego seleccione la casilla junto a "Desconocidos".

(**)Para los dispositivos con Android 4.0 o superior, vaya a "Ajustes▶ Seguridad" o "Configuración▶ Seguridad y Bloqueo de pantalla "y luego seleccione la casilla junto a" Desconocidos "y confirme su elección.

Paso 2: Conecte el ordenador y el dispositivo a misma red WiFi.
App Inventor le mostrará automáticamente la aplicación que está construyendo, pero sólo si su equipo (que se ejecuta la aplicación Inventor) y su dispositivo Android (que ejecuta el MIT EA2 App Companion), están conectados a la misma red WiFi.

Paso 3: Abra un proyecto de App Inventor y conéctelo a su dispositivo. Vaya a la url de App Inventor y abra un proyecto (o cree uno nuevo, yendo a *Proyecto▶ Iniciar Nuevo proyecto* y asigne a su proyecto un nombre).
A continuación, seleccione "**Connect**" y "**Al Companion**" en el menú superior en el navegador EA2:

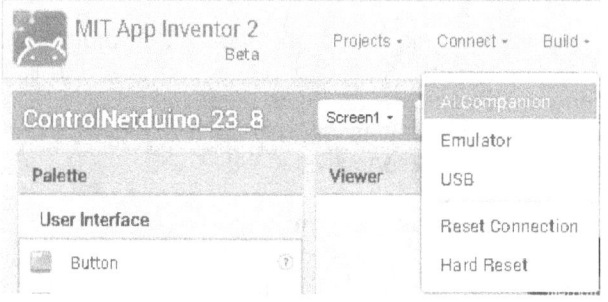

Se mostrará en su navegador una ventana emergente con un código QR. En el Smartphone o tableta, inicie la aplicación MIT App Companion tal y como lo haría con cualquier aplicación. Luego haga clic en el botón "Escanear código QR", y escanee el código que aparece en la ventana de la aplicación Inventor:

Netduino 2 en Español

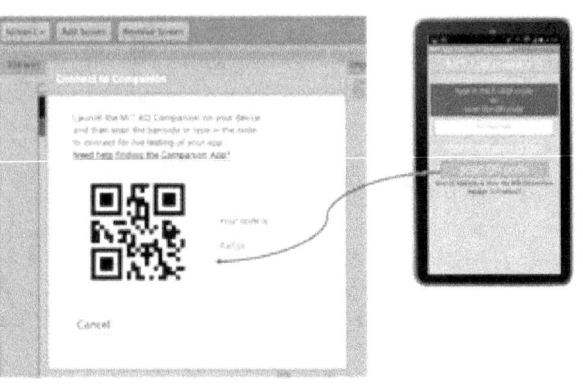

Si tiene problemas para escanear el código QR o el dispositivo no tiene un escáner, escriba el código que aparece en el ordenador en el área de texto del MIT App Companion en su dispositivo Android exactamente como se muestra (el código está directamente debajo del código QR donde pone "**Su código es**" y consta de seis caracteres).

En su terminal, en la aplicación, eleja el botón naranja "**Conectar con el código**" y escriba los seis caracteres. Tenga cuidado con escribir un retorno de carro o "Enter": sólo debe escribir los seis caracteres seguidos.

A los pocos segundos, en ambos casos, debería ver la aplicación que está construyendo en su dispositivo, actualizándose ésta a medida que realice cambios en su diseño y bloques, gracias a una característica llamada "pruebas en vivo".

1.2-Programación con un emulador

Si usted no tiene un Smartphone o tableta Android, aún puede construir aplicaciones con App Inventor, pues App Inventor proporciona un emulador de Android, que funciona de modo similar a un terminal Android, aunque lógicamente con algunas limitaciones.

El emulador aparece en la misma pantalla de su ordenador, de modo que usted puede probar sus aplicaciones en éste, y aún distribuir la aplicación a los demás, incluso a través de la Play Store.

Esta modalidad es muy usada en formación, que se desarrolla principalmente en los emuladores, proporcionando sólo al final terminales Android para la prueba final.

125

Para utilizar el emulador, primero tendrá que instalarlo en el equipo (esto no es necesario para la solución wifi) y a continuación iniciar el emulador.

A continuación se detallan los pasos para hacerlo:

Paso 1: *Instalación del paquete de software de instalación de App Inventor.*
Antes de empezar, debe saber que debe realizar la instalación desde una cuenta que tenga privilegios de administrador, y si ha instalado una versión anterior de App Inventor, tendrá que desinstalar ésta antes de instalar la versión más reciente.

Si usted elige utilizar el cable USB para conectar a un dispositivo, entonces usted también tendrá que **instalar los controladores de Windows** para su Smartphone Android.

Se recomienda siga los siguientes pasos:
- Descargue el instalador apropiado al sistema operativo de su ordenador. En caso de que su equipo tenga Windows como S.O., vaya a http://appinv.us/aisetup_windows.
- Busque el archivo AppInventor_Setup_Installer_v_2_1.exe (~ 101 MB) en su carpeta de descargas o el escritorio. La ubicación de la descarga en el equipo depende de la configuración de su navegador.
- Abra el archivo.
- Haga clic a través de los pasos del instalador. No cambie la ubicación de la instalación. El directorio será diferente dependiendo de su versión de Windows, y si está validado en su ordenador como administrador.
- Es posible que se le pregunte si desea permitir que un programa de un editor desconocido haga cambios en el equipo. Haga clic en Sí.
- Para localizar el software de configuración en la mayoría de los casos, App Inventor debe ser capaz de localizar el software de instalación por su cuenta. Si está utilizando una máquina de 64 bits, debe escribir **Archivos de programa (x86)** en lugar de **Archivos de programa.** Además, si usted no ha instalado el software como administrador, (por defecto la instalación será en **C: \ Archivos de programa)**, usted tendrá que buscar la ruta correcta.

Paso 2:*Uso de aiStarter*
El emulador requiere el uso de un programa llamado *aiStarter* que

permitirá al navegador comunicarse con el emulador. Este programa se instaló al ejecutar el paquete de instalación de App Inventor, por lo que se iniciará automáticamente al iniciar sesión en su cuenta ejecutándose de forma invisible en segundo plano

Habrá accesos directos a **aiStarter** desde el escritorio, en el **menú Inicio** y en "**Todos los programas**". Para iniciar aiStarter, si aún no lo está, haga doble clic en su icono: usted sabrá que ha lanzado con éxito aiStarter cuando vea una ventana del símbolo de sistema ejecutando en segundo plano con el título "aiStarter".

Paso 3: *Abrir un proyecto App Inventor y conéxion al emulador.*
En primer lugar, vaya a App Inventor y abra un proyecto (o cree uno nuevo: **Proyecto►** *Iniciar Nuevo proyecto* y ponga a su proyecto un nombre).

Luego, desde el menú de App Inventor (recuerde que el software basado en la nube App está en la url **ai2.appinventor.mit.edu**), vaya al **menú Conectar** y haga clic en la opción **emulador**.

Usted obtendrá un aviso diciendo que el emulador está conectando. Tenga en cuenta que iniciar el emulador puede tardar un par de minutos.

El emulador inicialmente aparecerá con una pantalla en negro vacío. Espere hasta que el emulador esté listo, con un fondo de pantalla de color.

Después de que aparezca el fondo, deberá esperar hasta que el terminal emulado haya terminado de preparar su tarjeta SD (habrá un aviso en la parte superior de la pantalla del terminal mientras se está preparando la tarjeta). Una vez conectado, el emulador se iniciará y mostrará la aplicación que tiene abierta en App Inventor.

2-Pasos para crear una aplicación

Gracias al framework App inventor, se pueden crear aplicaciones móviles sin conocimientos de programación Java, pues éste permite que la programación móvil sea sencilla y esté al alcance de cualquiera.

Casi cualquier tipo **de aplicación se puede crear con App Inventor** (todo depende de las ganas y empeño que le ponga), por ejemplo p**uede crear una aplicación Android para interactuar con Netduino 2 Plus** (como vamos a ver en el capítulo final del libro), y cuando este lista, subirla a Google Play y, además, obtener ingresos directos si la aplicación la pone de pago o incluso incluye publicidad usando los principales métodos de monetización como pueden ser **Admob, Leadbolt o Mobpartner.**

Para crear una aplicación con Mit App Inventor se deben seguir tres pasos:

Paso1: Diseño interfaz

El *diseño "estético" de la aplicación*, es la fase de creación de una aplicación en la que se **seleccionan los componentes visuales para su aplicación** (botones, cajas de texto, imágenes, etc.).

Los componentes son los elementos básicos que utilizamos para hacer aplicaciones Android (algo así como los "ingredientes de una receta").

Algunos componentes son muy simples, como una Etiqueta (**Label**), que sólo muestra el texto en pantalla, o un botón (**Button**) que se pulsa para iniciar una acción.

Otros componentes son más elaborados: un lienzo (**Canvas**) que puede almacenar imágenes fijas o animaciones, un sensor de movimiento (**AccelerometorSensor**) que funciona como un mando de Wii y detecta cuando se mueve o agita el Smartphone, los componentes para hacer o enviar mensajes de texto, los componentes para reproducir música y vídeo, los componentes para obtener información de sitios Web (que emplearemos intensivamente en nuestra aplicación que interactúa con Netduino), etc.

Para utilizar un componente en su aplicación, debe hacer clic, arrastrar y soltar en el visor (**Viewer**) en el centro de **Designer**. Cuando se agrega un componente al visor, también aparecerá en la lista de componentes a la derecha de éste.

Netduino 2 en Español

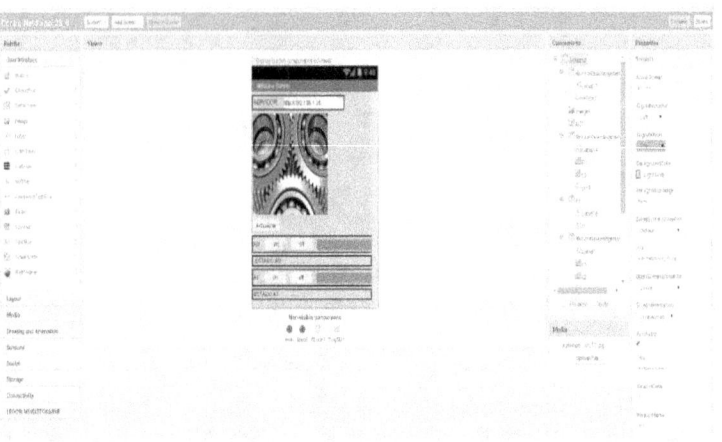

Los componentes se encuentran en el lado izquierdo de la ventana del diseñador, con el título **Palette,** y son los elementos básicos que se utilizan para construir aplicaciones para Android.

Para utilizar un componente en su aplicación, usted tendrá que hacer clic y arrastrarlo hacia la parte central del **Diseñador**. Cuando se agrega un componente al **Visor,** también aparecerá en la lista de componentes (**Components**) en el lado derecho del visor.

Los componentes tienen propiedades (**Properties)** que se pueden ajustar para cambiar la forma en que el componente aparece dentro de la aplicación. Para ver y cambiar las propiedades de un componente, primero debe seleccionar el componente deseado en la lista de componentes.

Los valores de algunas propiedades de ciertos componentes no son modificables, aunque aún así, si se podrán consultar. Los componentes tienen propiedades que pueden cambiar la forma en que éstos aparecen o actúan dentro de la aplicación.

Para ver y cambiar las propiedades de un componente, primero debe seleccionar el componente deseado en la lista de componentes.

Paso 2: Adición de la lógica

El *editor de bloques*, es la parte de App Inventor donde se escogen **los bloques lógicos,** que definen cómo se comportará la aplicación, estableciendo lo que los componentes deben hacer y cuándo hacerlo (por ejemplo, que debe ocurrir cuando el usuario pulsa un botón).

129

El Editor de bloques se ejecuta en una ventana independiente, pinchando en la solapa **Blocks** .Tiene dos fichas en la esquina superior izquierda: **Built-In** (incorporados) **y My Blocks** (mis bloques). Los botones de debajo de cada ficha se amplían y muestran los bloques cuando se hace clic.

Los bloques Built-In son el conjunto estándar de bloques que están disponibles para cualquier aplicación que construya. Los bloques mostrados al pulsar **My Blocks,** contienen bloques específicos que están vinculados al conjunto de componentes que haya elegido para su aplicación.

Utilice el **Editor de bloques** para ensamblar los bloques que definen el comportamiento de los componentes, como por ejemplo puede usar bloques **when..do** para definir controladores de eventos (que le dicen a los componentes lo que deben hacer cuando sucede algo), usar **bucles** para repetir acciones, emplear **variables** para almacenar valores usar **listas** para almacenar valores de una misma tipología accesibles por un índice, utilizar **persistencia** para guardar datos en el teléfono, etc.

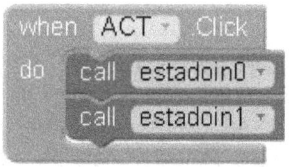

A continuación, se van a citar **los bloques básicos** de la programación del MIT App Inventor, que luego vamos a utilizar con una aplicación real que interactuará con Netduino 2+ (y que por cierto se puede descargar).

EVENTOS

Las aplicaciones creadas con Mit con App Inventor son **claramente orientadas a eventos**, pues no llevan a cabo un conjunto de instrucciones en un orden predeterminado secuencial (como ocurre en la mayoría de los lenguajes procedurales), sino que su funcionamiento se basa en reaccionar ante los eventos que pueden ser automáticos (por ejemplo, cuando se sobrepasa un límite) o iniciados por los propios usuarios (por ejemplo, al pulsar un botón).

En App Inventor, la respuesta ante eventos se hace directamente

mediante código, para indicar acciones que respondan ante un conjunto de eventos-respuestas, dentro de los bloques **when**... **do**.

Por ejemplo, en este bloque que usaremos en la aplicación que interactúa con Netduino 2+, tratamos de las acciones a realizar cuando pulsemos el botón llamado **b2**, que gestionaremos mediante el bloque **when b2.Click do**, desencadenando una serie de acciones que irán después del **do,** como son componer una url, hacer una petición http y leer el resultado obtenido, inicializar una variable global con la respuesta (llamada **valorpin0**) y por último llamar a una subrutina gráfica (llamada **pintar**).

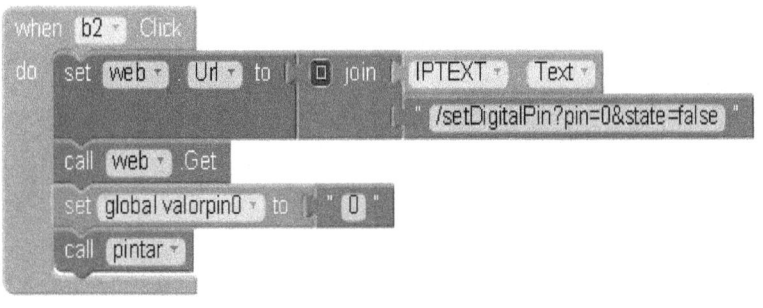

CONDICIONALES

Con App Inventor, se pueden programar controladores de eventos tal y como hemos visto anteriormente, donde veíamos como una aplicación debe responder, por ejemplo, cuando el usuario hace clic en un botón, pero, a veces, una aplicación no debe responder a un evento al mismo tiempo para todos los casos. En esos casos, es donde entrarían en acción los bloques conocidos como "**condicionales**", los cuales permiten a la aplicación tomar decisiones ante determinadas circunstancias, formando esto la base de lo que se conoce como "inteligencia artificial".

En el siguiente ejemplo, cuyo bloque usaremos en la aplicación que interactúa con Netduino 2+, vamos a ver lo que hace la rutina **colorpin0**, la cual colorea la caja asociada al **pin0** en función de la información disponible

Primero inspeccionamos el valor de la variable global **pin0**. Si ese valor es igual a "1", mediante el bloque **if get global pin0="1" then**, entonces y solo en ese caso, la aplicación pintará la caja asociada al **pin0** de color verde y se escribirá en la etiqueta el texto "activo ". En caso contrario

(bloque ".. **else ...**"), si el valor de la variable no vale "1", pintaremos de color rojo la caja asociada al **pin0** y se escribirá el texto "apagado".

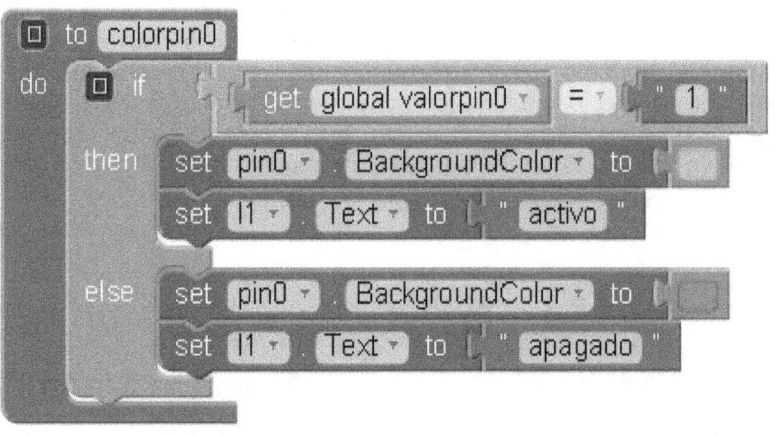

VARIABLES

Los valores de las variables en App Inventor son *células de memoria* temporales donde podemos almacenar valores de todo tipo, que al cerrar el programa perderán su contenido.Por ejemplo vamos a usar este tipo de células de memoria para guardar el valor del estado de un pin de Netduino 2+, actualizando su valor en función del estado de los pines físicos de Netduino 2+.

En el siguiente bloque, que usaremos en la aplicación que interactúa con Netduino 2+, vamos a ver cómo reaccionar ante la pulsación de dos botones diferentes (corresponden al botón **c2** de apagar y el botón **c1** de encender) cambiando el estado del pin correspondiente de Netduino 2+ mediante una orden vía http, y asignando seguidamente el nuevo valor a la variable que almacena su valor.

Este primer bloque, que ya hemos visto y que usaremos en la aplicación que interactúa con Netduino 2+, se centra en las acciones a realizar cuando pulsemos el botón llamado **c2**, que finalmente llevaran a activar a nivel bajo el pin1.

Al pulsar éste botón, desencadenaremos una serie de acciones, como serán componer una url que dará la orden a Netduino 2+ para que tome el valor lógico alto el **pin1**, leer a continuación el resultado de la petición y finalmente asignar el valor de la variable global (llamada **valorpin0**) al

Netduino 2 en Español

valor lógico "0" (es decir apagado) mediante el bloque **set global valorpin1 to "0"**.

Para terminar la ejecución de este bloque, se llama a una subrutina llamada pintar (**call pintar**), que se encargará de refrescar el color asociado a la caja que describe el estado real de los pines de Netduino 2+, que pretendemos controlar desde nuestra aplicación.

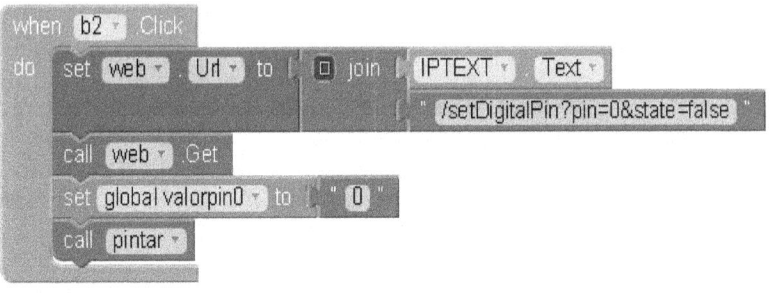

Ahora veamos otro segundo bloque, similar al bloque anterior y que usaremos también en la aplicación que interactúa con Netduino 2+.

Cuando pulsemos el botón llamado **c1** (de "on" para el pin1), también desencadenara una serie de acciones similares **b2** , como son componer una url y lanzarla hacia Netduino 2+ para que tome el valor lógico "1" al pin1 de éste , inicializando el valor de la variable global (llamada **valorpin0**) al valor lógico "1" (a diferencia del bloque anterior que se asignaba al valor lógico "0") mediante el bloque **set global valorpin1 to "1"**, y por último, al igual que en el bloque anterior, también llamando a una subrutina (llamada **pintar**) para refrescar el color asociado a ese pin1.

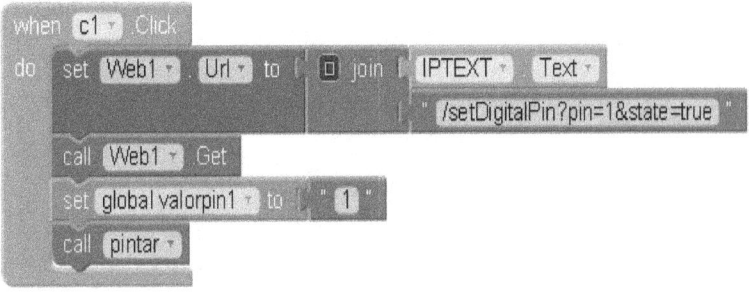

133

LISTAS

Las listas de datos se encuentran en muchas de las aplicaciones (de hecho cuando utiliza Facebook, por ejemplo, hay una lista de sus amigos, una lista de sus actualizaciones de estado, etc.), de modo que, con casi cualquier software desarrollado para Android, es probable que también haya listas de datos implicados (por ejemplo, en su aplicación, usted también puede realizar un seguimiento de sus amigos, los números de Smartphone, una lista de sus cuentas del pasado en un juego, etc.).

App Inventor, como la mayoría de los lenguajes de programación, ofrece también una manera de manejar listas permitiendo procesar los elementos de una lista, de modo que para realizar la misma operación en cada elemento con App Inventor, normalmente se utilizará una operación para cada bloque en una lista.

En el siguiente ejemplo, que usaremos en la aplicación que interactúa con Netduino 2+, vamos a ver como procesar el resultado de una petición http analizando el resultado en XML que nos devolverá Netduino 2+, usando también una lista llamada "**ítems**".

Con la línea **initialize local ítems to create empty list,** crearemos una lista vacía llamada **ítems,** que puede contener un número indeterminado de elementos.

Ahora, para cada elemento que obtengamos del parseo de procesar el fichero XML, que exploraremos gracias al subbloque **for each ..in list ..do,** añadiremos un nuevo elemento a la lista mediante el subbloque **add ítems to list.**

En este bloque, después de insertar los elementos en la lista **ítems,** recorremos la lista **ítem** mediante el segundo subbloque **for each ..in list ..do.** En este caso, buscaremos entre la lista el tag **pin1,** en cuyo caso su valor lo asignaremos a la variable **global valorpin1.**

Netduino 2 en Español

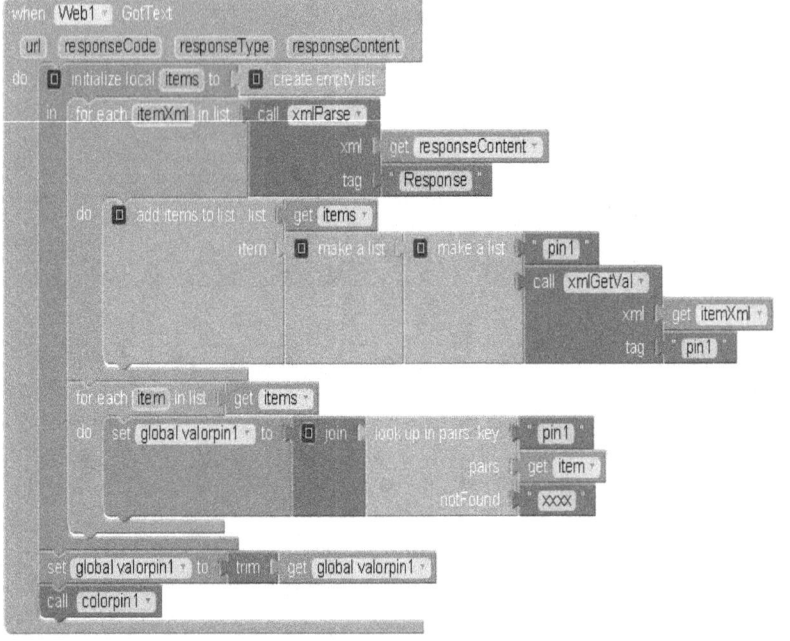

PERSISTENCIA

Al ingresar información en una página o en otra aplicación, la información se suele guardar en su terminal de modo que la próxima vez que usted visita esa página o ejecuta esa aplicación, suele también tener el mismo valor, es decir, se trata entonces de datos persistentes. Lo contrario, como es el caso de los datos almacenados en variables, sería que los datos "mueren" cuando se cierra la aplicación (por eso se llaman datos transitorios)

Vemos como los datos persistentes se almacenan en un sistema de base de datos o archivo, y no sólo en la memoria a corto plazo de la aplicación. Para ofrecer persistencia, App Inventor proporciona el componente **TinyDB** para el almacenamiento de datos en el propio sistema de archivos del Smartphone o tableta de forma persistente.

En el siguiente ejemplo , en este nuevo bloque que usaremos también en la aplicación que interactúa con Netduino 2+ , almacenaremos una variable persistente que llamaremos **iptext,** mediante el procedimiento **TinyDB1.StoreValue**, el cual recuperará el valor **IPTEXT** y lo salvará en la variable **iptext** para guardarla en un fichero del Smartphone.

135

En el segundo caso, recuperaremos una variable persistente llamada **iptext** , mediante el procedimiento **TinyDB1.GetValue**, el cual recupera el valor **iptext** almacenado en un fichero del teléfono. Es importante el parámetro **valueIfTagNotThere,** que asigna un valor por defecto en caso de estar vacío o de no encontrarlo.

PROCEDIMIENTOS

Refactorizar significa modificar el código para que sea más legible y fácil de mantener (por ejemplo, la extracción de código duplicado es una forma común de refactorizar), pero realmente desde la perspectiva de un programador, refactorizar no cambia el comportamiento de la aplicación (aunque la hace más fácil de entender y por tanto de mantener).

La idea básica es que el software cambia mucho, pero cuando se realizan cambios en el software, usted no quiere tener que encontrar, y también cambiar, un montón de "dependencias", por ejemplo, en otro código que haga lo mismo.

Es mejor tener código que haga una cosa en particular en un único lugar llamado **procedimiento** o **función**, y llamar a éste desde todos los lugares que lo necesiten: entonces, si el procedimiento tiene que ser cambiado, se cambia en un sólo lugar.

Netduino 2 en Español

Así surge la idea de los procedimientos que básicamente son secuencias con nombre de instrucciones (bloques) que pueden llamarse como un todo.

En App Inventor, puede definirse un procedimiento, colocar bloques en él, y luego se puede llamar desde cualquier parte de la aplicación.

En el siguiente procedimiento, que veremos en la aplicación que interactúa con Netduino 2+ llamado **colorpin0**, analizamos el valor de la variable global **valorpin0,** comprobando si es un valor alto, en cuyo caso pintaríamos la caja de **pin0** en verde y pondríamos en la caja **l1** "activo" o, en caso contrario, pintaríamos de color rojo la caja **pin0** y pondremos en la caja **l1** "apagado".

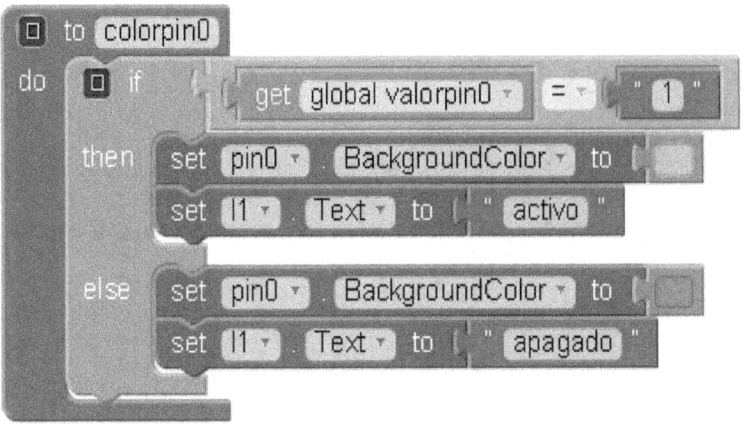

Este otro procedimiento llamado **colorpin1**, muy similar al anterior, también sin argumentos, se analiza ahora el valor de la variable global **valorpin1**, comprobando si es un valor alto, en cuyo caso pintaremos la caja de **pin1** en verde y pondremos en la caja **l2** "activo", o en caso contrario, pintaremos de color rojo la caja **pin1** y pondremos en la caja **l2** "apagado".

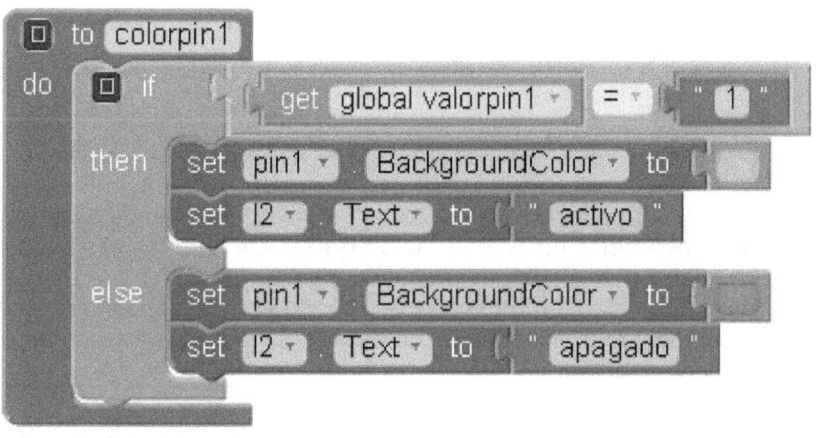

Finalmente, aquí tenemos la potencia de la refactorización: un nuevo procedimiento sin argumentos llamado "**pintar**", que llama precisamente a los dos procedimientos anteriores (refrescan en **pin0, pin1, l1 y l2** el estado de los pines 0 y 1 de Netduino 2+ obtenidos de inspeccionar las variables globales **valorpin0** y **valorpin1**).

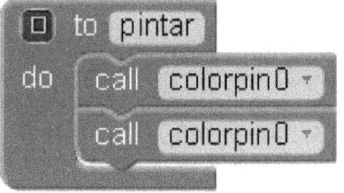

PASO3: DESPLIEGUE

Una vez diseñado el interfaz y dotado de lógica, es momento de desplegar la aplicación para su prueba en el terminal o en un emulador. Si se está utilizando un dispositivo Android y tiene una conexión wifi a Internet, puede simplemente capturar el código QR generado por el

software de APP Inventor con la **opción Build ▶ App (provide QR code for .apk)** mediante un lector de QR y aceptar la instalación de este en su Smartphone (de hecho esta es la opción es la que recomienda encarecidamente el MIT).

En caso de usar un emulador, usted tendrá que instalar un software en su ordenador, como hemos visto al principio de este capítulo, para que pueda utilizar el emulador de Android en la pantalla, tal y como hemos en el apartado anterior.

15 Aplicación móvil que interactúa con N2+

1-Introducción

A continuación vamos a desarrollar la creación de una pequeña aplicación con MIT App Inventor que será operativa al 100%, y nos permitirá interactuar con Netduino 2+ por medio de nuestro Smartphone Android sobre dos salidas binarias digitales de la placa.

Antes de empezar con el desarrollo de la aplicación, vamos a definir nuestro entorno que estará compuesto por un Netduino 2+ conectado a la misma red que emplearemos en nuestro programa, al que conectaremos los pines **D0** y **D1** respectivamente a dos ledes (o cualquier carga mediante el circuito adecuado), y sobre el que instalaremos el servidor **NeonMika Web-Server** explicado en el capítulo 13.

Respecto a la conexión de los ledes podemos hacerlo directamente dado que se hará sólo a efectos de pruebas, conectando los pines 0 y 1 respectivamente a los cátodos de los diodos ledes y ambos ánodos al terminal 0V de Netduino 2+, como se muestra en la imagen siguiente.

Carlos Rodríguez Navarro

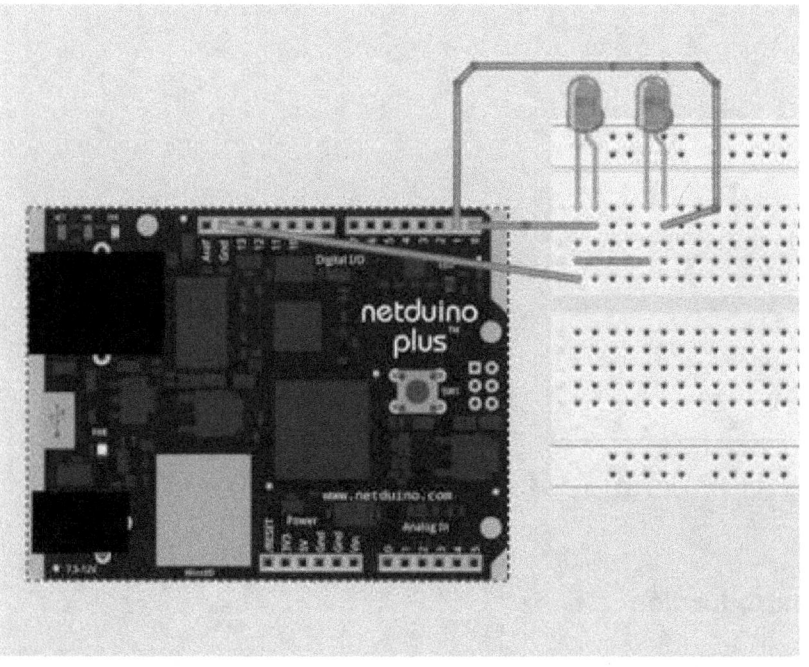

NeonMika.Webserver, como hemos visto, es un servidor web pre-configurado, muy fácil de extender, que necesita una mínima (o ninguna) línea de código pudiendo lograr grandes resultados para controlar su Netduino 2+.

Para instalar NeonMika.Webserver en Netduino 2+ sólo tiene que seguir los siguientes pasos:

1. Descargar el código de aquí
   http://neonmikawebserver.codeplex.com/downloads/get/785664
   (necesitará registrarse previamente en codeplex).
2. Descomprimir el fichero en un directorio de su ordenador.
3. Ejecutar Visual Studio.
4. Cargar el fichero de solución que está en la ruta de
   "Executables" (por ejemplo
   ..\neonmikawebserverV1.2\Executeable\NeonMikaWebserver\Ne
   onMiIkaWebServer.slh)
5. Implementar el proyecto a su Netduino 2+.

Una vez tengamos conectados los dos ledes a los pines **D0** y **D1** de Netduino 2+ y éste **cuente con conectividad** vía Ethernet (si no se dispone de conexión por cable al router se puede conectar vía PLC's o

142

con un repetidor wifi), lo primero que debemos saber es conocer la dirección IP asignada a Netduino 2+ (a no ser que haya configurado en apartado de red con una Ip fija).

Para **averiguar la dirección IP** con la que nos comunicaremos con el servidor web, bien podemos forzar a Netduino 2+ a que tome una determinada dirección IP, o bien podemos investigar cual es la dirección IP servida por el servidor DHCP.

En ambos casos, deberemos usar la herramienta incluida en todas las versiones de Microsoft.NET Framework llamada **MFDeploy,** la cual suele estar incluida dentro del directorio del Framework en Tools (por ejemplo en v4.3 suele estar en "C:\Program Files (x86)\Microsoft .NET Micro Framework\v4.3\Tools\MFDeploy.exe").

Una vez lanzado la herramienta **Mdfdeploy**, nos iremos a **Device**, seleccionando USB y en ese momento, nos debería aparecer, si tenemos conectado Netduino 2+ a nuestro ordenador, en el combo de la derecha, un ítem correspondiente a la versión de Netduino 2+.

Para averiguar la dirección Ip que está utilizando Netduino 2+, simplemente nos iremos al Menú **Target** y pincharemos en *Device capabilities*, y en seguida veremos en la primera línea, la dirección Ip de Netduino 2+.

Obtenida la dirección IP de Netduino 2+, asegurándonos que el servidor web se está ejecutando correctamente (led azul interno parpadeando a ráfagas lentas) y tiene conexión de red, suponiendo que la dirección IP de Netduino 2+ sea 192.168.1.35, a efectos de pruebas podemos hacer las siguientes peticiones http directamente desde un navegador:

- Obtener el estado lógico del pin0:
  http://192.168.1.35 /getDigitalPinState?pin=0.
- Obtener el estado lógico del pin1
  http://192.168.1.35 /getDigitalPinState?pin=1.
- Forzar a nivel bajo el pin0:
  http://192.168.1.35 /setDigitalPin?pin=0&state=false.
- Forzar a nivel alto el pin0:
  http://192.168.1.35 / setDigitalPin?pin=0&state=true.
- Forzar a nivel bajo el pin1;
  http://192.168.1.35 /setDigitalPin?pin=1&state=false.
- Forzar a nivel alto el pin1:
  http://192.168.1.35 / setDigitalPin?pin=1&state=true.

Un vez comprobado que el servidor web está funcionando y responde a todas las peticiones http anteriores, vamos a desarrollar una aplicación para la plataforma Android, la cual enviará peticiones http a Netduino 2+, y procesara la salida XML devuelta por Netduino 2+ para permitirnos conocer el estado de los pines, así como cambiar su estado de un forma sencilla y gráfica todo de forma remota y desde un terminal Android..

2- Diseño interfaz gráfico

Para empezar nuestro proyecto, arrancaremos Mit App Inventor, creando un nuevo proyecto.

En modo **Designer** crearemos 5 layouts:

1- Añadiremos una etiqueta (label), donde colocaremos el texto fijo "SERVIDOR" y una caja de texto (textbox) que llamaremos **IPTEXT** donde introduciremos la dirección IP de Netduino 2+. Asimismo, en este, se ha añadido una imagen de 200x200 pixels y un botón (button) que hemos renombrado como **ACT**.

2- Añadiremos una etiqueta (label), donde colocaremos el texto fijo "A0" y una caja de texto (textbox) que llamaremos **pin0,** cuyo color de fondo cambiaremos según el estado del pin **A0**. Asimismo, se complemente con dos botones (button) **b1** y **b2**, en el que incluiremos el texto "on" y "off" respectivamente.

3- Añadiremos una etiqueta (label), donde colocaremos el texto fijo "ESTADO A0" y una caja de texto (textbox) que llamaremos **l1**, donde a introduciremos el estado del pin0.

4- Añadiremos una etiqueta (label), donde colocaremos el texto fijo "A1" (label1) y una caja de texto (textbox) que llamaremos **pin1**, cuyo color de fondo cambiaremos según el estado del pin A1. Asimismo, se complementa con dos botones (button) **c1** y **c2**, en el que incluiremos el texto "on" y "off" respectivamente.

5- Añadiremos una etiqueta (label), donde colocaremos el texto fijo "ESTADO A1", y una caja de texto (textbox), que llamaremos **l2**, donde introduciremos el estado del pin1.

Asimismo, fuera de los layouts, arrastraremos a la zona no visible los componentes:

- **Web1.**
- **Web.**
- **Tinidb1.**

De este modo, el aspecto de la pantalla quedará de este modo:

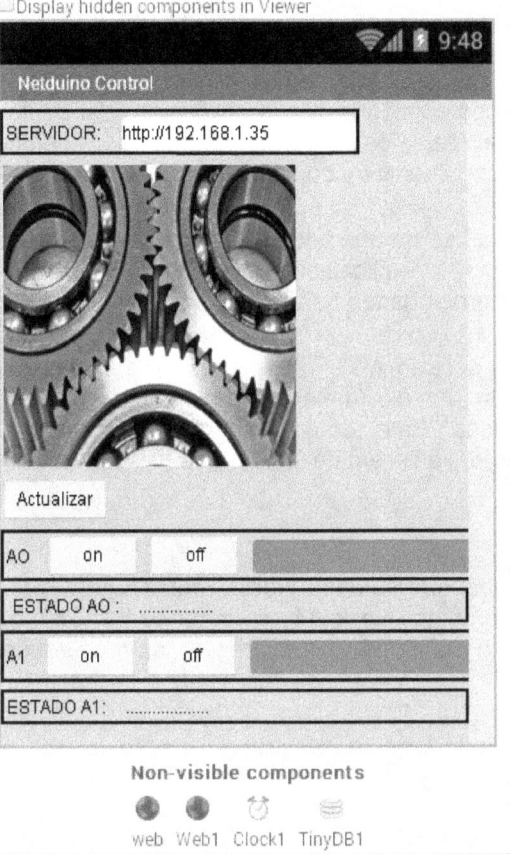

De forma pormenorizada, vamos a ver las propiedades de los diferentes elementos que componen el interfaz.

El primer elemento es el objeto **screeen1,** donde podemos especificar el título de la ventana, la versión, el color de fondo o incluso una posible imagen de fondo:

En el layout horizontal **Horizontalarragement17,** incluiremos **label17** y la caja de texto **IPTEXT**. Se han mantenido los valores de las propiedades por defecto.

En la siguiente etiqueta **label17**, mostraremos el texto **Servidor** en color negro, dejando todo lo demás en las opciones por defecto, (dimensionamiento automático y fuentes y color predefinidos).

El siguiente elemento es la caja de texto llamada **IPTEXT,** que contendrá el valor de la dirección IP del servidor web. Obsérvese que se inicializa a un valor por defecto, que como hemos visto corresponde a la IP de Netduino, y que podemos configurar con la aplicación MDFdeploy. El resto de opciones, las dejamos en los valores por defecto (dimensionamiento automático y fuentes y color predefinidos).

Para mejorar la estética global, se ha añadido una imagen en el interfaz llamado **image1**, para hacer la aplicación más atractiva. Se han mantenido los valores de las propiedades por defecto.

Se ha añadido un botón con el texto **ACT** (de Actualizar), dejando todo lo demás en las opciones por defecto (dimensionamiento automático y fuentes y color predefinidos). Este botón servirá para forzar la actualización del interfaz con el valor real de los pines 0 y 1 de Netduino 2+.

Ahora veremos el layout horizontal llamado **HorizontalArragement14,** en el que se incluirán los elementos **Label14**, los botones **b1** y **b2**, así como la etiqueta **pin0**. Se han mantenido los valores de las propiedades por defecto.

En la siguiente etiqueta, llamada **label14,** mostraremos el texto "A0" en color negro, dejando todo lo demás en las opciones por defecto (dimensionamiento automático y fuentes y color predefinidos).

Se ha añadido un botón llamado **B1** con el texto "ON", dejando todo lo demás en las opciones por defecto (dimensionamiento automático y fuentes y color predefinidos). Este botón servirá para forzar a nivel alto el **pin0** de Netduino 2+.

Se ha añadido un botón llamado **b2** con el texto "OFF", dejando todo lo demás en las opciones por defecto, (dimensionamiento automático y fuentes y color predefinidos). Este botón servirá para forzar a nivel bajo el **pin0** de Netduino 2+.

Netduino 2 en Español

El siguiente elemento llamado **pin0,** es la caja de texto que contendrá el valor del **pin0** de Netduino 2+ representado por un color rojo (off), verde (on) o gris (sin sincronizar). Obsérvese que se inicializa al atributo **backgroud** a un valor gris claro. El resto de opciones, la dejamos en los valores por defecto (dimensionamiento automático y fuentes y color predefinidos).

Ahora veremos un layout horizontal llamado **h1,** en el que se incluirán los elementos **Label16** y **l1**. Se han puesto las dimensiones automáticas.

En la siguiente etiqueta llamada **label16,** mostraremos el texto "ESTADO A0"en color negro, dejando todo lo demás en las opciones por defecto, (dimensionamiento automático, con fuentes y color predefinidos).

En la siguiente etiqueta llamada **l1,** mostraremos el texto "........" en color negro, para ser reemplazado por la lógica con el valor real del estado del **pin0**. Se ha dejando todo lo demás en las opciones por defecto (dimensionamiento automático con fuentes y color predefinidos).

Ahora veremos el layout horizontal llamado **HorizontalArragement2,** en el que se incluirán los elementos **Label1**, los botones **c1** y **c2**, así como la etiqueta **pin1**. Se han mantenido los valores de las propiedades por defecto.

En la siguiente etiqueta llamada **Label1,** mostraremos el texto "**A1"** en color negro, dejando todo lo demás en las opciones por defecto (dimensionamiento automático con fuentes y color predefinidos).

Se ha añadido un nuevo botón llamado **c1,** con el texto "ON", dejando todo lo demás en las opciones por defecto (dimensionamiento automático con fuentes y color predefinidos). Este botón servirá para forzar a nivel bajo el pin 1 de Netduino 2+.

Se ha añadido un botón llamado **c2,** con el texto "OFF", dejando todo lo demás en las opciones por defecto (dimensionamiento automático con fuentes y color predefinidos). Este botón servirá para forzar a nivel bajo el **pin1** de Netduino 2+.

El siguiente elemento llamado **pin1,** es la caja de texto que contendrá un valor de **pin1** representado por un color rojo (off), verde (on) o gris (sin sincronizar). Obsérvese que se inicializa al atributo **backgroud** a un

valor gris claro. El resto de opciones, las dejamos en los valores por defecto (dimensionamiento automático con fuentes y color predefinidos).

Ahora veremos el layout horizontal llamado **HorizontalArragement16,** en el que se Incluirán los elementos **Label15** y **I2**. Se han mantenido los valores de las propiedades por defecto.

En la siguiente etiqueta llamada **Label15,** mostraremos el texto "**ESTADO A1**", en color negro, dejando todo lo demás en las opciones por defecto (dimensionamiento automático con fuentes y color predefinidos).

En la siguiente etiqueta llamada **I2**, mostraremos el texto "**.......**" en color negro para ser reemplazado por la lógica con el valor real del estado del **pin1**, dejando todo lo demás en las opciones por defecto (dimensionamiento automático con fuentes y color predefinidos).

Por último, tenemos tres componentes no visibles, el primero es un elemento web llamado **web,** en el que hemos dejado todo lo demás en las opciones por defecto.

También hemos definido un segundo elemento web llamado **web1,** en el que hemos dejado también todo lo demás en las opciones por defecto.

Por último, tenemos el elemento **TiniDB1**, que nos permite la persistencia de datos en nuestra aplicación.

3- Bloques lógicos de la aplicación

Una vez que tenemos diseñado el interfaz gráfico, antes de empezar con toda la lógica de la aplicación, dado que el resultado de las peticiones http que lanzaremos a Netduino 2+ será en formato **XML**, se hace necesario crear un procedimiento que sea capaz de procesar el formato de salida, ya que MIT App Inventor no incluye los métodos nativos para procesar o leer dicho formato.

El procedimiento principal es "**xmlParse**" que requiere dos parámetros: **XML** (corresponde a la cadena devuelta por Netduino en formato XML) y **tag** (la cadena a buscar para averiguar su valor devuelto en XML por Netduino 2+).

La idea de este procedimiento es recorrer la salida XML buscando primero la cadena "**<tag>**" donde "tag" corresponde al valor del pin de

Netduino 2+, para después capturar todo lo que venga después hasta la cadena "**</**", pues ese será el valor que nos interesa recuperar.

Por ejemplo si enviamos la petición a Netduino 2+:
http://192.168.1.35/getDigit_alPinState?pin=0, ésta podría ser la respuesta:

> Respuesta
> This XML file does not appear to have any style information associated with it. The document tree is shown below.
> <Response>
> <pin0>1</pin0>
> </Response>

Evidentemente todo este contenido iría en la cadena XML y el tag seria "pin0", y lo que nos interesaría procesar seria sólo el valor 1, pues va entre la subcadena "**<pin0>**" y la subcadena "**</pin0>**".

Por último, el procedimiento salvará este resultado en el elemento 1 de la lista **tmp**, guardando todo lo que venga después en el elemento 2 de la citada lista.

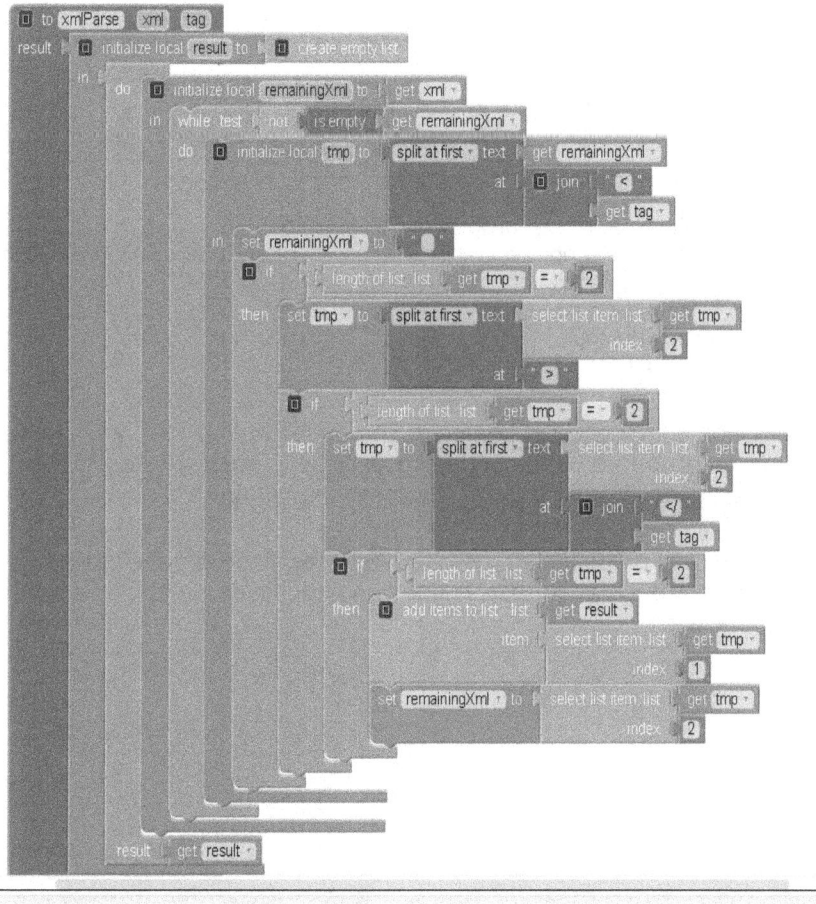

Visto la esencial del procedimiento que analizara el XML devuelto de Netduino 2+, ahora vamos a ver como lo invocamos gracias al procedimiento **xmlGetVal**, el cual requiere la cadena XML devuelta de la respuesta de Netduino 2+ (que pasaremos al procedimiento anterior) y el tag, que corresponderá al valor de la etiqueta a buscar en la cadena.

El funcionamiento, como vemos, se traduce en llamar al procedimiento anterior **xmlParse** comentado anteriormente con los argumentos *XML* y tag, recuperando el ítem 1 de la lista **tmp** y devolviendo finalmente este valor.

Netduino 2 en Español

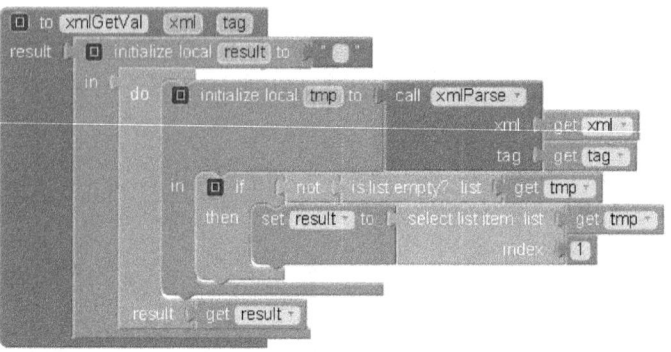

Visto el código encargado de procesar la salida en XML que nos devuelve Netduino 2+, vamos a ver ahora el resto de bloques que constituirán la lógica de la aplicación.

En primer lugar definimos dos variables globales, en las que almacenaremos el estado lógico de los pines 0 y 1 de nuestro Netduino 2+:

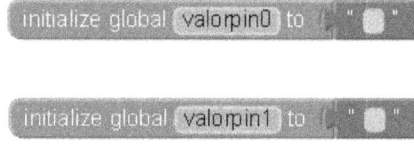

En el bloque de código de más abajo, se captura lo que se introduzca en la caja **IPText** llamando al procedimiento **TinyDB1.StoreValue**, para que lo almacene persistentemente en la variable **iptext** de forma permanente, gracias al componente **TiniDB1**.

El siguiente bloque, corresponde a un nuevo procedimiento sin parámetros, cuya misión es actualizar la caja **IPText** con el valor almacenado de forma permanente en la variable persistente **iptext**. Esta funcionalidad, la conseguiremos llamando al procedimiento **TinyDB1.Getvalue,** para que recupere el valor de la variable **iptext** almacenada de forma permanente.

Carlos Rodríguez Navarro

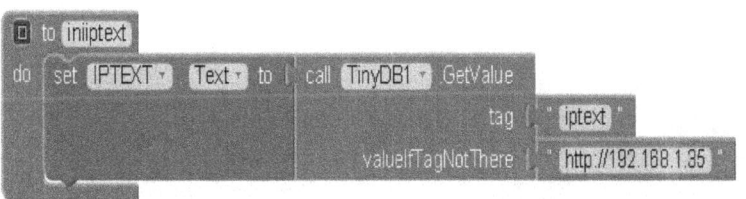

Al iniciar la aplicación, se llama en primer lugar al evento **Screen .Initialize**, donde podemos anidar las comprobaciones y acciones iniciales que deseemos.

En nuestro caso, en primer lugar, necesitamos saber el valor de la dirección IP del servidor de Netduino 2+, que obtendremos gracias a la llamada al procedimiento **iniiptext**.

Después, una vez que ya sabemos la dirección IP del servidor, llamaremos a los procedimientos **estadoin0** y **estadoin1,** los cuales se encargaran de obtener el valor real lógico de los pines 0 y 1 de Netduino.

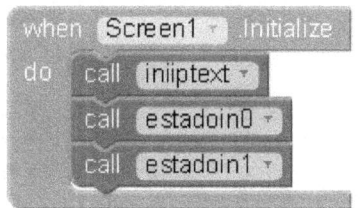

Mediante el siguiente procedimiento, llamado **estaodoin0,** lanzaremos una petición http a Netduino 2+ para averiguar el estado del pin0.

Obsérvese que para obtener la url, unimos el valor de la caja **IPTEX** (donde tenemos la dirección IP del servidor) y la cadena **"/getsDigitalPinState?pin=0**", que corresponde al método de NeonMika Web-Server para comprobar el estado lógico del pin 0.

152

Netduino 2 en Español

Mediante el procedimiento llamado **estaodoin1,** similar al anterior, lanzaremos una petición http a Netduino 2+ para averiguar el estado del pin1.

Obsérvese que para obtener la url, unimos el valor de la caja **IPTEX** (donde tenemos la IP del servidor) y la cadena "**/getsDigitalPinState?pin=1**", que corresponde al método de NeonMika Web Server, para comprobar el estado lógico del pin 1.

El siguiente procedimiento, llamado **colorpin0,** analiza el valor de la variable global **valorpin0,** comprobando si es un valor alto, en cuyo caso pintaremos la caja de **pin0** en verde, y pondremos en la caja **I1** "activo," o en caso contrario, pintaremos de color rojo la caja **pin0** y pondremos en la caja **I1** "apagado".

153

Carlos Rodríguez Navarro

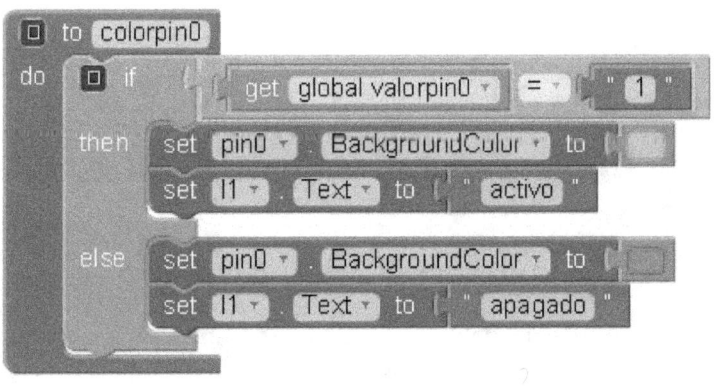

El siguiente procedimiento, llamado **colorpin1**, muy similar al anterior también sin argumentos, y analiza el valor de la variable global **valorpin1,** comprobando si es un valor alto, en cuyo caso pintaremos la caja de **pin1** en verde y pondremos en la caja **I2** "activo," o en caso contrario, pintaremos de color rojo la caja **pin1** y pondremos en la caja **I2** "apagado".

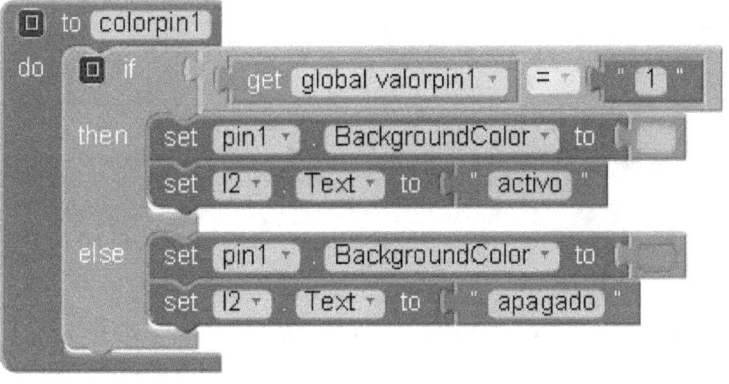

Finalmente, el procedimiento sin argumentos "**pintar**", llama precisamente a los dos procedimientos anteriores, cuya misión es refrescar los elementos **pin0**, **pin1**, **I1** y **I2** con el estado real de los pines 0 y 1 de Netduino, obtenido de inspeccionar las variables globales **valorpin0** y **valorpin1**.

Netduino 2 en Español

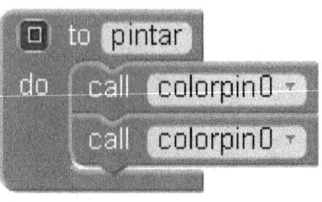

El siguiente bloque se activa si pulsamos el botón **ACT** (actualizar), el cual llama respectivamente a los procedimientos ya comentados **estadoin0** y **estadoin1**, los cuales como sabemos, se encargan de realizar sendas peticiones http para saber el estado real de los pines 0 y 1.

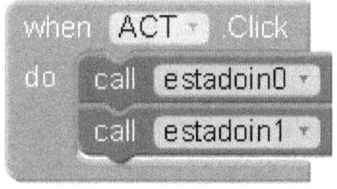

Los siguientes 4 bloques corresponden a la lógica asociada al pulsar los botones **on, off** de ambos pines, veamos con detalles el código de éstos:

El primer bloque, corresponde al botón **b2** que se asocia al estado "off" "del pin0, y en él componemos la petición http uniendo con un **join** la dirección IP del servidor con la cadena **"/setDigitalPin?pin0&state=false"** (que corresponde al método de NeonMika Web-Server para poner a estado lógico false al pin 0).

Una vez tenemos la cadena http, hacemos la petición **get**, actualizando el valor de la variable global **valorpin0** al valor **cero**, y llamando a la rutina pintar ya comentada, cuya misión es refrescar el valor de los elementos **pin0, pin1, I1** y **I2** con el estado real de los pines 0 y 1 de

155

Netduino, por medio de la inspección de las variables globales **valorpin0** y **valorpin1**.

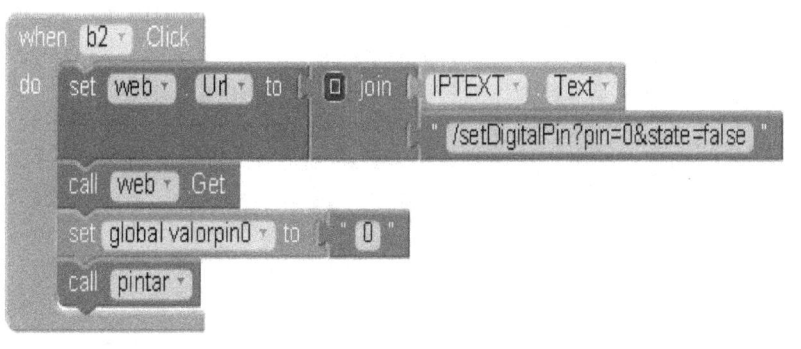

El segundo bloque, correspondiente al botón **b1**, se asocia al "on "del **pin0**, para lo cual primero componemos la petición http uniendo con un **join** la dirección IP del servidor con la cadena **"/setDigitalPin?pin0&state=true"** (corresponde al método de NeonMika Web Server para poner a estado lógico true al pin 0).

Una vez tenemos la cadena http, hacemos la petición **get**, actualizando el valor de la variable global **valorpin0** al valor **uno**, y llamando a la rutina pintar ya comentada, cuya misión es refrescar los elementos **pin0, pin1, l1** y **l2** con el estado de los pines 0 y 1 de Netduino 2+ por medio de la inspección de las variables globales **valorpin0** y **valorpin1**.

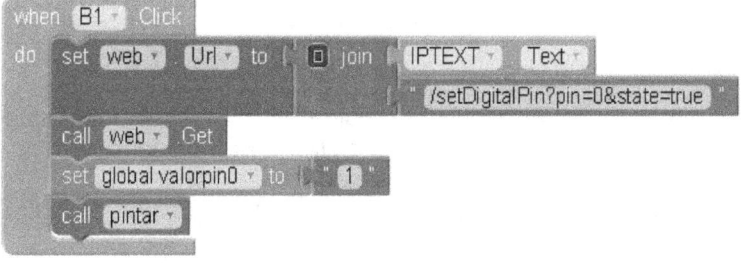

El tercer bloque, corresponde al botón **c2,** que se asocia al "off" del **pin1**, y en él componemos la petición http uniendo con un **join** la

dirección IP del servidor con la cadena **"/setDigitalPin?pin1&state =false**" (corresponde al método de NeonMika Web-Server para poner a estado lógico false al pin 1).

Una vez tenemos la cadena http, hacemos la *petición get*, actualizando el valor de la variable global **valorpin1** al valor **cero**, y llamando a la rutina **pintar** ya comentada, cuya misión es refrescar los elementos **pin0, pin1, I1** y **I2** con el estado de los pines 0 y 1 de Netduino 2+, por medio de la inspección de las variables globales **valorpin0** y **valorpin1**.

El cuarto bloque, corresponde al botón **c1**, se asocia al "on" del **pin1**, para lo cual primero componemos la petición http uniendo con un **join** la dirección IP del servidor con la cadena **"/setDigitalPin?pin1& state=true**" (corresponde al método de NeonMika Web-Server para poner a estado lógico true al pin1).

Una vez tenemos la cadena http, hacemos *la petición get*, actualizando el valor de la variable global **valorpin1** al valor **uno** lógico, y llamando a la rutina **pintar** ya comentada cuya misión es refrescar los elementos **pin0, pin1, I1** y **I2** con el estado de los pines 0 y 1 de Netduino 2+ por medio de la inspección de las variables globales **valorpin0** y **valorpin1**.

Para terminar, vamos a comentar los dos últimos bloques en esta aplicación que son los encargados de procesar la salida en XML de Netduino 2+ ante peticiones que lancemos hacia él, es decir, de analizar la respuesta que nos haya dado una petición http.

En este primer bloque, para evitar problemas de interacción con otros, sólo vamos a procesar las peticiones que tengan que ver con el primer **pin0** de Netduino, capturando la respuesta (responseCode) y almacenando ésta en la variable global **valorpin0,** y por último, actualizando el color de la caja reservada a este pin, así como el valor de su estado devuelto por Netduino 2+ en la respuesta.

Recordemos que si enviamos la petición a Netduino 2+ la petición: http://192.168.1.35/getDigitalPinState?pin=0 ,esta podría ser la respuesta:

Respuesta
This XML file does not appear to have any style information associated with it. The document tree is shown below.
<Response>
<pin0>1</pin0>
</Response>

Obsérvese que el primer contenido relevante de la salida obtenida de la petición http para procesar, corresponde al tag **"<Response>"** de la salida XML: por este motivo llamaremos al procedimiento **xmlParse** que buscará en toda la respuesta devuelta por Netduino 2+ esta cadena **"<Response>"** devolviendo como resultado todo lo que venga a continuación.

Una vez encontrado la subcadena, donde está el valor del estado real de Netduino 2+, llamaremos ahora al procedimiento **xmlGetVal** para procesar precisamente el tag **<pin0>**.

Ahora, como se ha explicado, llamando a **xmlGetVal** capturaríamos el valor asociado a **pin0,** dado que este valor está incluido en la cadena que va comprendida entre el tag <pin0>, y la siguiente subcadena "</" (esta cadena pues constituye el valor real del pin0).

Para terminar, como hemos procesado solo el valor **pin0**, salvaremos este en la variable **valorpin0.**

Netduino 2 en Español

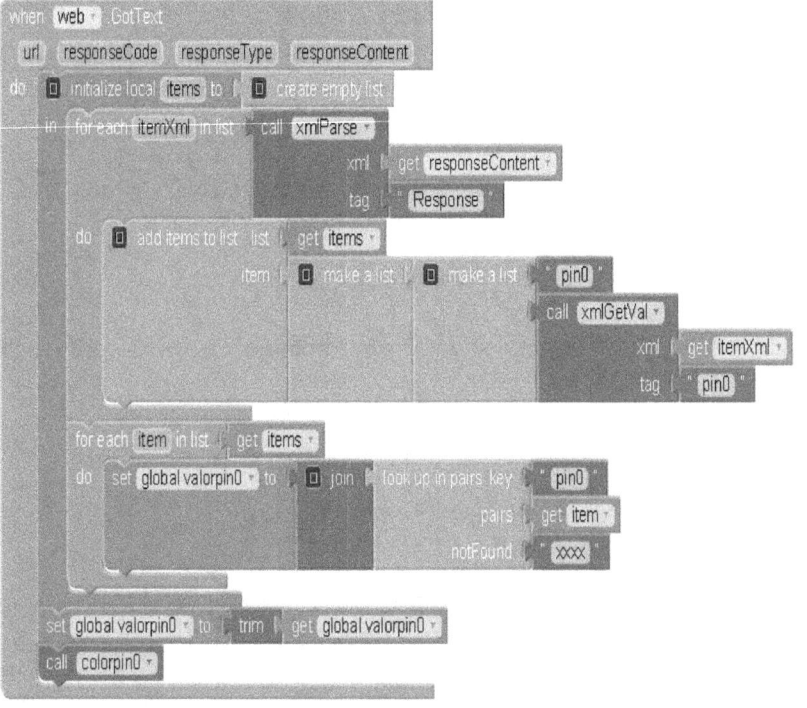

En el último bloque, para evitar problemas de interacción con el bloque anterior, sólo vamos a procesar las peticiones que tengan que ver con el primer **pin1** de Netduino 2+ capturando la respuesta (responseCode) de Netduino 2+ y almacenando la respuesta en la variable global **valorpin1,** y por último, actualizando el color de la caja reservada a este pin, así como el valor de su estado devuelto por Netduino 2+ en la respuesta.

Recordemos que si enviamos la petición a Netduino2 + la petición será http://192.168.1.35/getDigitalPinState?pin=1 ,y esta podría ser la respuesta:

Respuesta

This XML file does not appear to have any style information associated with it. The document tree is shown below.

<Response>
<pin1>0</pin1>
</Response>

Obsérvese, que el primer contenido relevante de la salida obtenida de la petición http para procesar, corresponde al tag **"<Response>"** de la salida XML. Por este motivo, llamaremos al procedimiento **xmlParse** que buscará en toda la respuesta devuelta por Netduino esta cadena **"<Response>"**, devolviendo como resultado todo lo que venga a

159

continuación.

Una vez encontrado la subcadena donde está el valor del estado real de Netduino 2+, llamamos ahora al procedimiento **xmlGetVal** para procesar precisamente el tag **<pin1>**.

Ahora ,como se han explicado, llamando a **xmlGetVal** capturaríamos el valor asociado a **pin1**, dado que este valor está incluido en la cadena que va comprendida entre el tag **<pin1>** y la siguiente subcadena "**</**" (esta cadena pues constituye el valor real del pin1).

Para terminar, salvaremos el valor **pin1** obtenido en la variable global **valorpin1**.

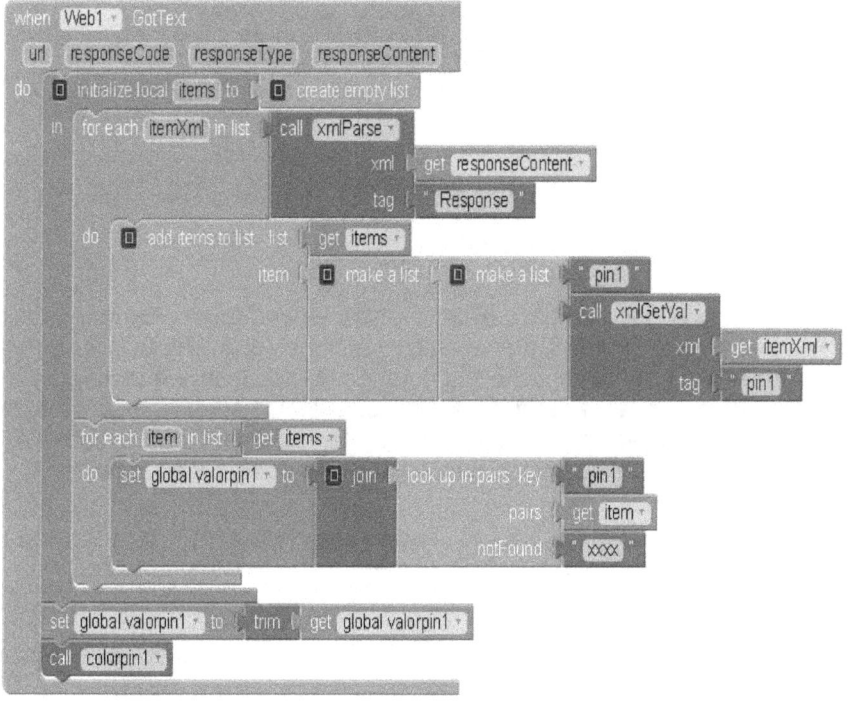

4- Aspecto final

Una vez concluido el diseño de los bloques lógicos, si no tenemos errores de sintaxis, solo se trata de lanzar esta aplicación a nuestro terminal mediante la opción **Build →App (provide QR code for .apk).**
Bien con un lector de códigos **QR** (como el lector BIDI) o bien con la propia aplicación del MIT App Inventor Companion, escanearemos el

código QR que se presenta en pantalla, con lo que se descargará éste en el terminal Android, comenzando a instalarse en cuanto lo aprobemos.

Una vez instalada la aplicación, aparecerá en el menú de aplicaciones como una aplicación más con el nombre "**ControlN**":

Teniendo en cuenta que en Netduino 2+ estamos ejecutando el Servidor NeonMika Server (debe destellar el led azul), si lanzamos la aplicación **ControlN,** siempre que haya conectividad entre Netduino 2+ y el terminal Android (es decir estén en la misma red wifi), veremos el interfaz que hemos diseñado con los estados del **pin0** y **pin1** con su valor real del Netduino 2+.

Tal y como hemos visto, si no hubiera conexión, habría error de conectividad y los botones se mostrarían en tonos grises.

En caso de haber dejado los pines activos (los dos ledes conectados a los pines 0 y 1 encendidos), la imagen que mostraría la aplicación sería algo similar a esta:

Carlos Rodríguez Navarro

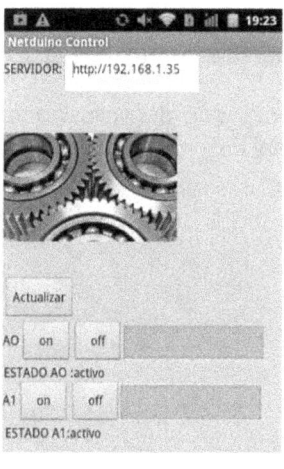

Ahora, si pulsamos el botón **off** de A1, se apagará el segundo led quedando el interfaz de A1 de color rojo y la etiqueta quedará con el texto de apagado.

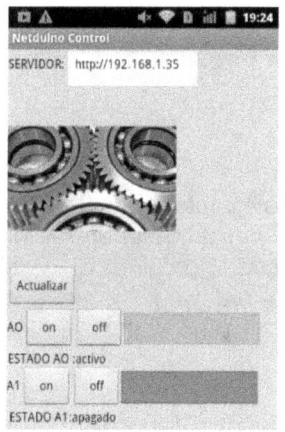

Si también pulsamos el botón **off** de A0, el interfaz de A0 aparecerá rojo, se apagará el primer led y la etiqueta de A0 mostrará el texto de apagado.

Netduino 2 en Español

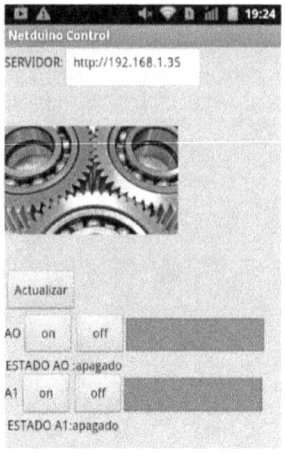

Nota: Puede descargar el código fuente de esta aplicación gratuitamente escaneando el siguiente código QR:

Asimismo, también puede descargar el código accediendo a la siguiente url:
http://gallery.appinventor.mit.edu/#page%3DApp%26uid%3D28356001%26label%3DNetduino%20Control%20%20by%20CRN

ANEXO1: Actualizar Netduino

Flashear siempre es un riesgo en casi todos los dispositivos programables, básicamente porque el 99,9% de las veces irá bien, y como fruto tendremos un sistema actualizado, pero desgraciadamente, puede haber un 0.1% donde no irá bien y el resultado no será agradable (en ingles lo llaman "bricked", o sea "ladrillo", pues un dispositivo con el firmware mal no sirve para nada, y por desgracia, no siempre se puede arreglar).

En el caso de contar con el Netduino con 4.1.0 (versión anterior) podría estar más que justificado realizar la actualización, pues el nuevo firmware 4.2.1 del que hemos tratado en este libro incluye las siguientes actualizaciones:

- Corrección de errores **PWM** (periodo / frecuencia ahora ajustada correctamente).
- 64Kb flash y 50% más de memoria RAM: 42KB (4.2.0.0) frente a 28KB (4.1.0.6).
- Visual Basic es un lenguaje soportado por **NETMF**.
- Nuevas clases **AnalogInput** y **PWM** básicos.
- Nuevos drivers **WinUSB**.
- Recolector de basura que recoge ahora plenamente objetos antiguos.

Para encontrar la versión actual de su firmware Netduino 2+, siga estos pasos:

1. Ir al inicio Menú▶ Programas▶ Microsoft. NET Micro Framework 4.2▶ Herramientas.
2. Ejecute **MFDeploy.exe**. Tenga cuidado al ejecutar MFDeploy.exe y no MFDeploy.exe.config (como las extensiones de archivo están ocultas de manera predeterminada no se verá siempre claro el ejecutable).
3. Conecte su Netduino 2+ a su PC usando un cable USB Micro.
4. En la sección de dispositivos en la parte superior, seleccione USB en lugar de serie. Su Netduino 2+ debe aparecer en la lista desplegable, si no fuese así, selecciónelo.
5. Seleccione la opción de menú Target, Capacidades de dispositivo.
6. En el cuadro de salida, encontrar el valor "SolutionReleaseInfo.solutionVersion". Esta es la versión del firmware.

Si su versión es la antigua (4.1.x), puede seguir bajo su responsabilidad, los tres siguientes pasos para actualizar el firmware:

1-Para empezar necesitará descargar e instalar la aplicación SAM-BA 2.12 CDC para Windows (XP, Vista, W7) desde la url http://www.atmel.com/tools/ATMELSAM-BAIN-SYSTEMPROGRAMMER.aspx (usted tendrá que registrarse para descargarla, pero no habrá ningún coste adicional).

Ahora para restablecer completamente su Netduino+, conecte con un cablecillo al pin de 3,3 V con el otro extremo la plaquita de metal que hay debajo del pin 0, mientras que el Netduino+ está encendido como se muestra en la ilustración.

Netduino 2 en Español

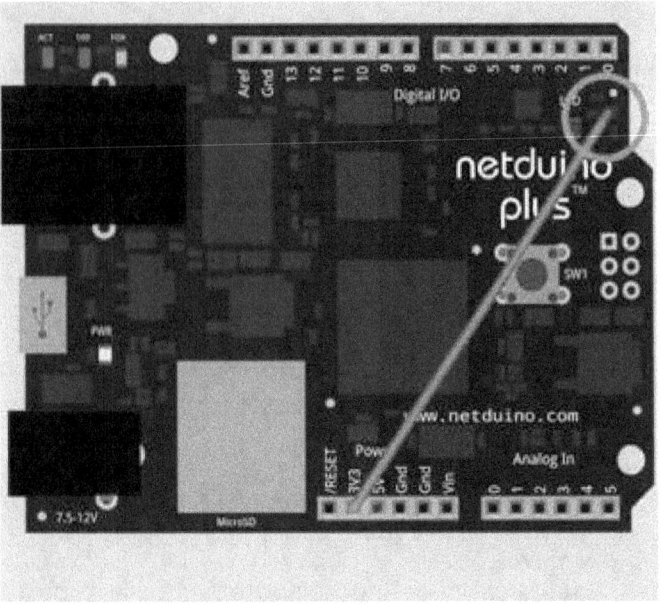

2-Ahora reconectar el Netduino+ al PC, de modo que se instalará un controlador para un puerto COM emulado. Una vez instalado el driver del puerto COM, inicie la herramienta SAM-BA (***).

(***)NOTA: Si, un controlador de un puerto COM emulado no está instalado en una máquina x64 de Microsoft ® Windows 7, pruebe lo siguiente: Pulse el botón de Windows y abra una ventana de Explorer. Haga clic en el botón "Propiedades del sistema": en este momento, usted encontrará un enlace al "Administrador de dispositivos" en la parte izquierda, haga clic en ella. Allí encontrará un dispositivo desconocido. Haga clic con botón derecho sobre él y seleccione "Actualizar software de controlador..." Ahora deje que busque automáticamente el driver (en realidad ejecutará Windows Update). Sabemos que no una cámara con detección de GPS pero como lo reconocerá con este nombre, como es un puerto COM, y SAM-BA v2.10_CDC puedo reconocerlo como tal, puede continuar con el reflasheo de su Netduino.

Seleccione el puerto serie (puerto COM) emulado del Netduino+ al que se conectará (normalmente será el com10), seleccionando el chip adecuado para su Netduino+ (el identificador del chip del procesador Atmel debería ser el "AT91SAM7X512-ek.", aunque el valor real dependerá del Netduino particular que usted tenga) y finalmente haga clic en "Conectar".

Carlos Rodríguez Navarro

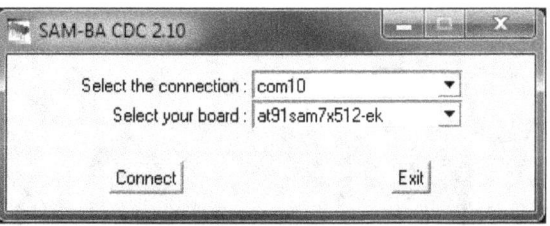

Ahora tenemos que ejecutar dos scripts:

1. En primer lugar, seleccione "**Boot from flash (GPNVM2)**" y haga clic en "**Ejecutar**".
2. Luego tenemos que hacer lo mismo con el script "**Enable Flash access**".

Ahora asegúrese de que estamos en la pestaña "**Flash**". En la caja "**Send File Name**", seleccione el archivo "**TinyBooterDecompressor.bin**" y haga clic en "**Enviar Archivo**".

Este archivo se encuentra en el ultimo post de este hilo:"http://forums.Netduino.com/index.php?/topic/5582-Netduino-plus-firmware-v420-update-1/ "

Entonces nos hará un pregunta sobre regiones bloqueadas. Haga clic en "**No**".

168

Netduino 2 en Español

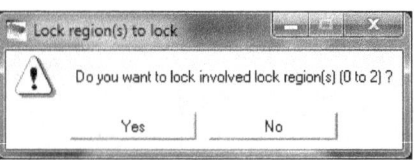

Tras unos instantes, ya habrá quedado instalado el **TinyBooterDecompressor,** de modo que podemos cerrar la herramienta SAM-BA y podremos comprobar si se ha instalado correctamente, para lo cual vuelva a conectar su Netduino y ejecute **MFDeploy.exe.** Seleccione USB y el modelo de su Netduino, a continuación, haga clic en **Ping.** Si responde, será señal de que el Netduino+ vuelve a estar pre operativo, y ya estará listo para grabar el firmware.

3-Ahora desde el MFDeploy haga clic en "Browse" y seleccione tanto el archivo **ER_FLASH** como el fichero **ER_CONFIG** que previamente haya descargado del último post de esta url:
http://forums.Netduino.com/index.php?/topic/5582-Netduino-plus-firmware-v420-update-1/. (Mantenga presionada la tecla "Ctrl" para seleccionar ambos archivos).

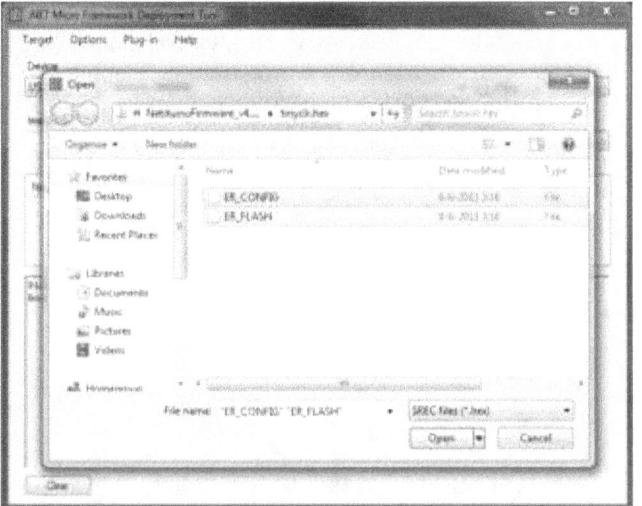

Una vez seleccionado ambos archivos, haga clic en **DEPLOY** (Implementar) .Ahora el proceso de flasheo pude tomar unos 4 o 5 minutos máximos…

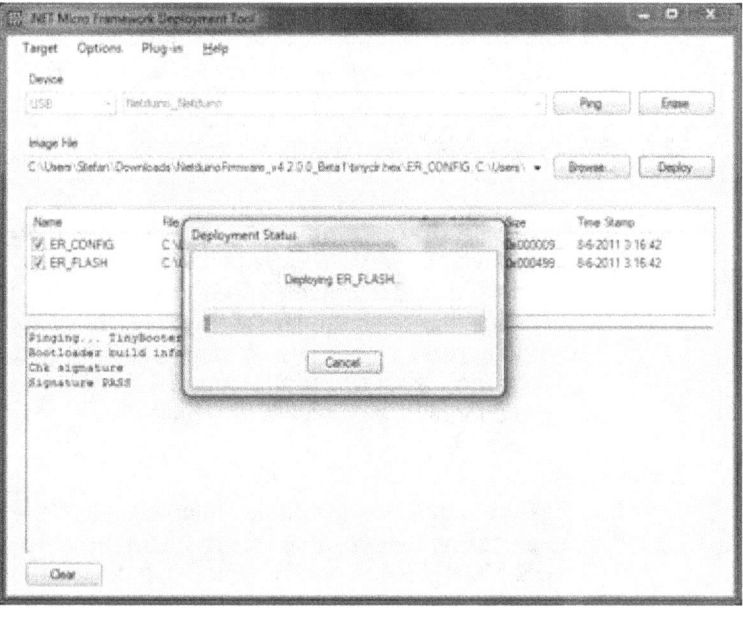

Una vez acabado el proceso, veremos el mensaje "**SIGNATURE PASS**", lo cual nos indica que se pude retirar el conector micro-USB y volver a conectarlo a su PC para completar la operación de flasheo.

Después de reiniciar el Netduino+, ahora puede comprobar la versión del firmware actual ejecutando otra vez el programa MdfDeploy seleccionando la opción de menú "**Target► Device Capabilities**".

Netduino 2 en Español

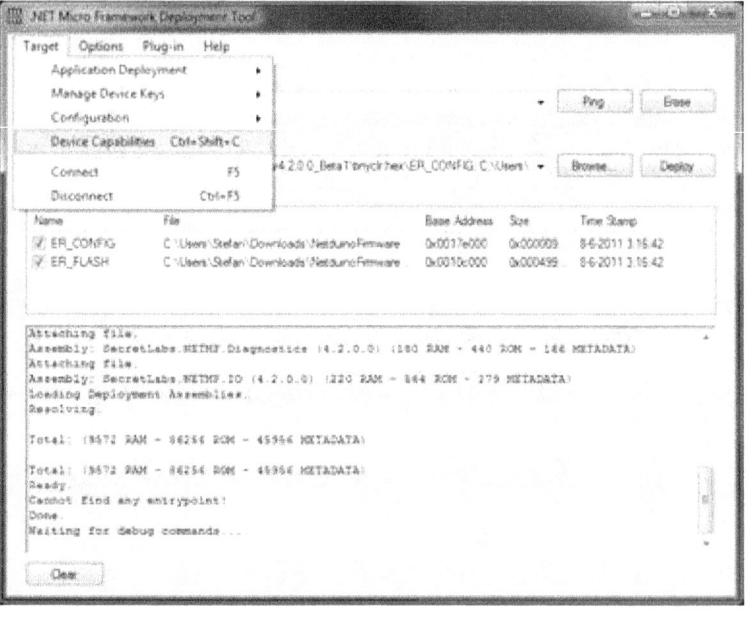

Esto sería un ejemplo de resultado:

HalSystemInfo.halVersion: 4.2.0.0
HalSystemInfo.halVendorInfo: Netduino Plus (v4.2.0.1) by Secret Labs LLC
HalSystemInfo.oemCode: 34
HalSystemInfo.modelCode: 177
HalSystemInfo.skuCode: 4097
HalSystemInfo.moduleSerialNumber: 00000000000000000000000000000000
HalSystemInfo.systemSerialNumber: 0000000000000000
ClrInfo.clrVersion: 4.2.0.0
ClrInfo.clrVendorInfo: Netduino Plus (v4.2.0.1) by Secret Labs LLC
ClrInfo.targetFrameworkVersion: 4.2.0.0
SolutionReleaseInfo.solutionVersion: 4.2.0.0
SolutionReleaseInfo.solutionVendorInfo: Netduino Plus (v4.2.0.1) by Secret Labs
LLC
SoftwareVersion.BuildDate: Sep 19 2012
SoftwareVersion.CompilerVersion: 410894
FloatingPoint: True
SourceLevelDebugging: True
ThreadCreateEx: True
LCD.Width: 0
LCD.Height: 0
LCD.BitsPerPixel: 0
AppDomains: True
ExceptionFilters: True
IncrementalDeployment: True
SoftReboot: True
Profiling: False
ProfilingAllocations: False
ProfilingCalls: False
IsUnknown: False

Restablecer configuración de red

Después de flashear un Netduino+ a la versión 4.2, debería también restablecer la configuración de red, pues con el flasheo perderá la dirección MAC de la tarjeta y todos los parámetros de red. Para restituir los valores de red con el firmware 4.2, es necesario mantener pulsado el botón de la placa Netduino mientras se conecta el USB, para poner el dispositivo en modo de arranque.

Ahora, lanzar **MFDeploy,** que se puede encontrar en el menú Inicio► Programas► Microsoft. NET Micro Framework"►**Herramientas.**

Seleccione USB y su modelo de Netduino+ desde el menú despegable.

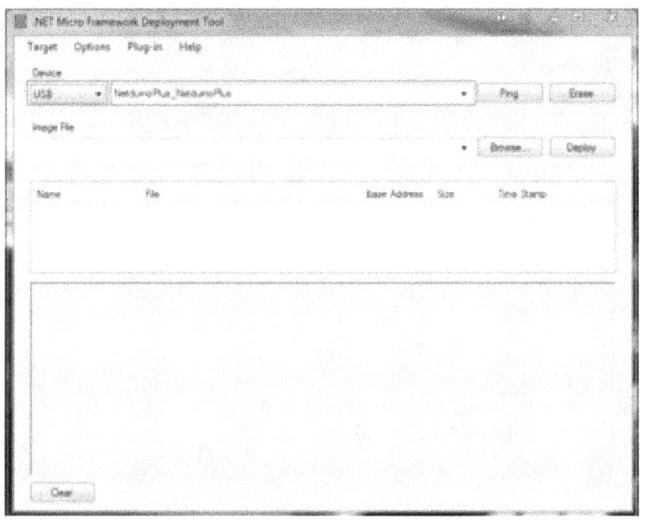

Pulse ahora el botón "**Ping**", para asegurarse de que su Netduino+ está funcionando correctamente, de modo que debería ver algo así como "Pinging... TinyCLR" después de un momento en la mitad inferior de la pantalla.

Ahora vaya desde el menú Target ► Configuration ► Network.

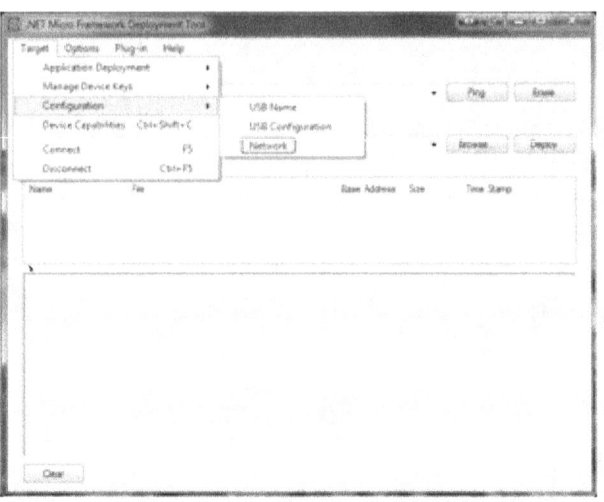

Esto abrirá la ventana de configuración de red, que contiene todos los valores de red de su Netduino+. Desde aquí se puede configurar una dirección IP estática o habilitar DHCP.

También puede configurar la dirección MAC y los ajustes de DNS para su Netduino+.

Lo más importante, es que restituya el valor de la dirección MAC que encontrará en la pegatina de la parte posterior de su placa Netduino+ y activar el DHCP (check Enable).

Network Configuration

Static IP address:	192.168.5.100
Subnet Mask:	255.255.255.0
Default Gateway:	192.168.5.1
MAC Address:	5c-86-4a-00-25-4b
DNS Primary Address:	8.8.8.8
DNS Secondary Address:	8.8.4.4
DHCP:	☑ Enable

Wireless Configuration

Authentication	None
Encryption	None
Radio	☐ 802.11a ☐ 802.11b ☐ 802.11g ☐ 802.11n
Encrypt Config Data	☐
Pass phrase	
Network Key	64-bit
ReKey Interval	
SSID	

Update   Cancel

Cambie al menos dichos ajustes, pulse **Update** (Actualizar) y la configuración de red de su Netduino+ se habrá realizado tras el reinicio de éste.

¡Respire, ya ha concluido el flasheo!

ANEXO 2: ERRORES

1-Error 0x8D131700 al compilar

El error 0x8D131700 al intentar compilar código para Netduino con Microsoft ® Visual Studio c#, es un error típico del propio compilador de Visual Studio y por tanto no achacable al código que estemos intentando compilar.
Podría tratar los siguientes dos pasos para solucionarlo:
1. Reinicie su PC. ¿Se puede compilar ahora?
2. Si falla el punto primero, instale todas las actualizaciones de Windows Update "críticos" e "importantes". Especialmente los que estén relacionados con todos los parches para .NET. ¿Se puede compilar ahora?

2-Cómo corregir el error 10060

Probablemente si está trabajando con **M2M** (machine to machine), y está enviando sockets a un servidor distante a pesar de que su programa en c# se compile y se ejecute aparentemente bien, puede obtener este error en ejecución con su Netduino+:

Exception System.Net.Sockets.SocketException – CLR_E_FAIL (1)
Message:
Microsoft.SPOT.Net.SocketNative::getaddrinfo [IP: 0000]
System.Net.Dns::GetHostEntry [IP: 0008]
System.Net.HttpWebRequest::EstablishConnection [IP: 00e1]
System.Net.HttpWebRequest::SubmitRequest [IP: 0013]
System.Net.HttpWebRequest::GetRequestStream [IP: 0008]
PachubeClient.Program::CreateRequest [IP: 0076]

```
#### PachubeClient.Program::Publish [IP: 0008] ####
#### PachubeClient.Program::Main [IP: 0052] ####
#### SocketException ErrorCode = 10060
#### SocketException ErrorCode = 10060
```

Pues bien, este error podría solucionarse reflasheando su Netduino+ a una versión superior, pero es mucho más fácil y seguro solucionarlo simplemente configurando la dirección IP de su Netduino + de forma estática.

Para hacerlo solo hay que lanzar el MDFDeploy ("C:\Program Files\Microsoft .NET Micro Framework\v4.2\Tools\MFDeploy.exe") y teniendo Netduino+ conectado por USB, y seleccionado en device USB y a su derecha **NetduinoPlus_NetduinoPlus** ir a Target►Configuration►Network.

Tras unos instantes aparecerá la ventana de red que deberemos dejar de un modo parecido a esta pantalla (des-chequeado DNS y cumplimentado el resto de campos).

3-Qué hacer si no se puede compilar su código

En ocasiones debido a continuos reflasheos de Netduino+, es posible que llegue un momento en que nos sea imposible compilar nuestro código dando error de comunicaciones, error desconocido, etc.

Para solucionar el problema en el 99,9% de las ocasiones tan sólo hay que ejecutar la utilidad **Mdfdeploy** incluida en el SDK (normalmente estará en C:\Program Files (x86)\Microsoft .NET Micro Framework\v4.3\Tools\MdfDeploy.exe).
En primer lugar pulsaremos en **Device** seleccionado USB ya que la opción por defecto es serial.

Acto seguido, si tenemos pinchado Netduino al USB, debería aparecer nuestro Netduino en la caja de la derecha de USB.

Para comprobar que Netduino está funcionando podemos pulsar **Ping** a lo que Netduino+ debería responder con "TinyCLR" y el valor de la IP Address.

Como vemos que responde, podemos proceder a eliminar el ultimo software que se compiló el cual probablemente será el origen del problema, para lo cual simplemente pulsaremos el botón"**Erase**".

Netduino 2 en Español

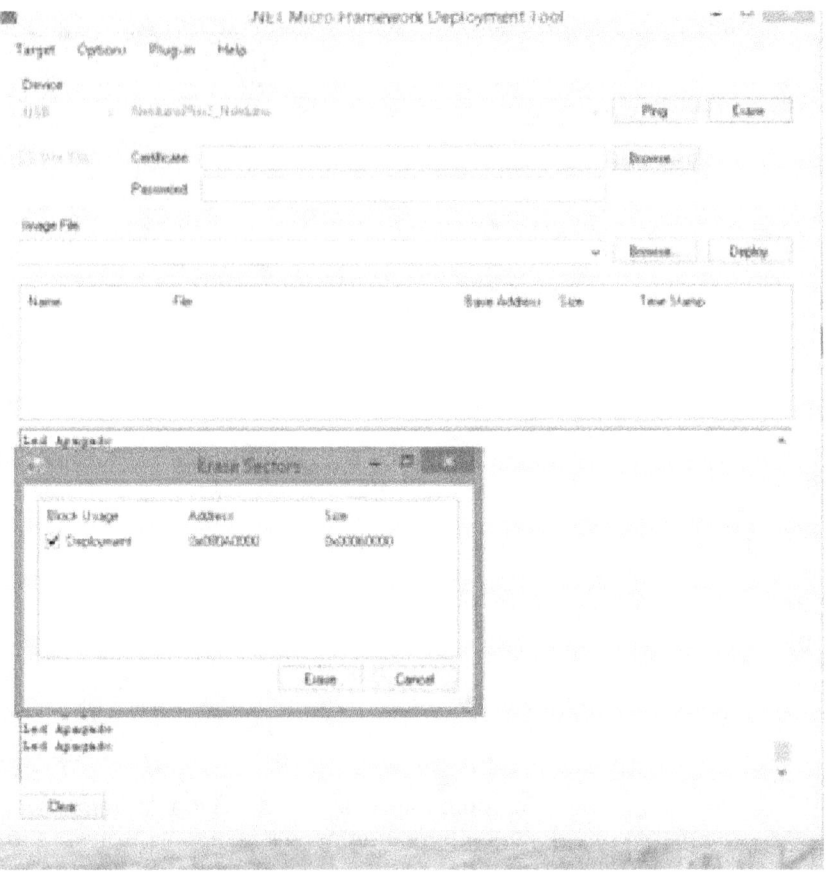

Nos pedirá confirmación por lo que deberemos volver a pulsar **Erase** (o Cancel si no estamos seguros).

Ahora nos irá apareciendo una barra de progreso hasta terminar al 100%, momento en el cual ya habremos eliminado el software de Netduino+, y por lo tanto, podemos volver a intentar compilar con el código que daba problemas.

ANEXO 3

Ejemplo de salida por consola al compilar el primer ejemplo "Hello World" del capítulo 3:

Found debugger!
Create TS.
Loading start at 154280, end 16c5dc
Assembly: mscorlib (4.2.0.0)
Assembly: Microsoft.SPOT.Native (4.2.0.0)
Assembly: Microsoft.SPOT.Hardware (4.2.0.0)
Assembly: Microsoft.SPOT.Net (4.2.0.0)
Assembly: System (4.2.0.0)
Assembly: Microsoft.SPOT.Hardware.SerialPort (4.2.0.0)
Assembly: Microsoft.SPOT.IO (4.2.0.0)
Assembly: System.IO (4.2.0.0)
Assembly: Microsoft.SPOT.Hardware.PWM (4.2.0.1)
Assembly: SecretLabs.NETMF.Diagnostics (4.2.0.0)
Loading Deployment Assemblies.
Attaching deployed file.
Assembly: SecretLabs.NETMF.Hardware.Netduino (4.2.1.0)
Attaching deployed file.
Assembly: Netduino.Application1 (1.0.0.0)
Attaching deployed file.
Assembly: SecretLabs.NETMF.Hardware (4.2.0.0)
Resolving.
The debugging target runtime is loading the application assemblies and starting execution.
Ready.
'Microsoft.SPOT.Debugger.CorDebug.dll' (Administrado): se cargó 'C:\Program Files (x86)\Microsoft .NET Micro Framework\v4.2\Assemblies\le\mscorlib.dll', símbolos cargados.
'Microsoft.SPOT.Debugger.CorDebug.dll' (Administrado): se cargó 'C:\Program Files (x86)\Microsoft .NET Micro Framework\v4.2\Assemblies\le\Microsoft.SPOT.Native.dll', símbolos cargados.
'Microsoft.SPOT.Debugger.CorDebug.dll' (Administrado): se cargó 'C:\Program Files (x86)\Microsoft .NET Micro Framework\v4.2\Assemblies\le\Microsoft.SPOT.Hardware.dll', símbolos cargados.
'Microsoft.SPOT.Debugger.CorDebug.dll' (Administrado): se cargó 'C:\Program Files (x86)\Microsoft .NET Micro Framework\v4.2\Assemblies\le\Microsoft.SPOT.Net.dll', símbolos cargados.
'Microsoft.SPOT.Debugger.CorDebug.dll' (Administrado): se cargó 'C:\Program Files (x86)\Microsoft .NET Micro Framework\v4.2\Assemblies\le\System.dll', símbolos cargados.
'Microsoft.SPOT.Debugger.CorDebug.dll' (Administrado): se cargó 'C:\Program Files (x86)\Microsoft .NET Micro Framework\v4.2\Assemblies\le\Microsoft.SPOT.Hardware.SerialPort.dll', símbolos cargados.
'Microsoft.SPOT.Debugger.CorDebug.dll' (Administrado): se cargó 'C:\Program Files (x86)\Microsoft .NET Micro Framework\v4.2\Assemblies\le\Microsoft.SPOT.IO.dll', símbolos cargados.
'Microsoft.SPOT.Debugger.CorDebug.dll' (Administrado): se cargó 'C:\Program Files (x86)\Microsoft .NET Micro Framework\v4.2\Assemblies\le\System.IO.dll', símbolos cargados.

'Microsoft.SPOT.Debugger.CorDebug.dll' (Administrado): se cargó 'C:\Program Files (x86)\Microsoft .NET Micro Framework\v4.2\Assemblies\le\Microsoft.SPOT.Hardware.PWM.dll', simbolos cargados.
'Microsoft.SPOT.Debugger.CorDebug.dll' (Administrado): se cargó 'C:\Program Files (x86)\Secret Labs\Netduino SDK\Assemblies\v4.2\le\SecretLabs.NETMF.Diagnostics.dll', símbolos cargados.
'Microsoft.SPOT.Debugger.CorDebug.dll' (Administrado): se cargó 'C:\Program Files (x86)\Secret Labs\Netduino SDK\Assemblies\v4.2\le\SecretLabs.NETMF.Hardware.Netduino.dll', símbolos cargados.
'Microsoft.SPOT.Debugger.CorDebug.dll' (Administrado): se cargó 'C:\Users\carlos\documents\visual Studio 2012\Projects\NetduinoApplication2\NetduinoApplication2\bin\Debug\le\NetduinoApplication2.exe', símbolos cargados.
'Microsoft.SPOT.Debugger.CorDebug.dll' (Administrado): se cargó 'C:\Program Files (x86)\Secret Labs\Netduino SDK\Assemblies\v4.2\le\SecretLabs.NETMF.Hardware.dll', símbolos cargados.
El subproceso '<Sin nombre>' (0x2) terminó con código 0 (0x0).
Led parpadeante
Led encendido
Led Apagado
Led encendido
Led Apagado
Led encendido
Led Apagado

ANEXO 4

Salida por la consola de Visual Studio ejecutando el programa NeonMika Web Server :

webserver:
Found debugger!
Create TS.
Loading start at 806a238, end 8085f74
Assembly: mscorlib (4.2.0.0)
Assembly: Microsoft.SPOT.Native (4.2.0.0)
Assembly: Microsoft.SPOT.Hardware (4.2.0.0)
Assembly: Microsoft.SPOT.Net (4.2.0.0)
Assembly: System (4.2.0.0)
Assembly: Microsoft.SPOT.Hardware.SerialPort (4.2.0.0)
Assembly: Microsoft.SPOT.IO (4.2.0.0)
Assembly: System.IO (4.2.0.0)
Assembly: Microsoft.SPOT.Hardware.PWM (4.2.0.1)
Assembly: Microsoft.SPOT.Hardware.Usb (4.2.0.0)
Assembly: SecretLabs.NETMF.Diagnostics (4.2.0.0)
Assembly: SecretLabs.NETMF.Hardware.Netduino (4.2.1.0)
Assembly: Microsoft.SPOT.Hardware.OneWire (4.2.0.0)
Assembly: Microsoft.SPOT.Time (4.2.0.0)
Loading Deployment Assemblies.
Attaching deployed file.
Assembly: JSONLib (1.0.0.0)
Attaching deployed file.
Assembly: SecretLabs.NETMF.Hardware.NetduinoPlus (4.2.2.0)
Attaching deployed file.
Assembly: NeonMika.Webserver (1.0.0.0)
Attaching deployed file.
Assembly: NeonMika.XML (1.0.0.0)
Attaching deployed file.
Assembly: SecretLabs.NETMF.Hardware.PWM (4.2.2.0)

Carlos Rodríguez Navarro

Attaching deployed file.
Assembly: NeonMikaWebserver (1.0.0.0)
Attaching deployed file.
Assembly: SecretLabs.NETMF.Hardware.AnalogInput (4.2.2.0)
Attaching deployed file.
Assembly: SecretLabs.NETMF.Hardware (4.2.0.0)
Attaching deployed file.
Assembly: NeonMika.Util (1.0.0.0)
Resolving.

The debugging target runtime is loading the application assemblies and starting execution.
Ready.
'Microsoft.SPOT.Debugger.CorDebug.dll' (Administrado): se cargó 'C:\Program Files
(x86)\Microsoft .NET Micro Framework\v4.2\Assemblies\le\mscorlib.dll', símbolos cargados.
'Microsoft.SPOT.Debugger.CorDebug.dll' (Administrado): se cargó 'C:\Program Files
(x86)\Microsoft .NET Micro Framework\v4.2\Assemblies\le\Microsoft.SPOT.Native.dll',
símbolos cargados.
'Microsoft.SPOT.Debugger.CorDebug.dll' (Administrado): se cargó 'C:\Program Files
(x86)\Microsoft .NET Micro Framework\v4.2\Assemblies\le\Microsoft.SPOT.Hardware.dll',
símbolos cargados.
'Microsoft.SPOT.Debugger.CorDebug.dll' (Administrado): se cargó 'C:\Program Files
(x86)\Microsoft .NET Micro Framework\v4.2\Assemblies\le\Microsoft.SPOT.Net.dll',
símbolos cargados.
'Microsoft.SPOT.Debugger.CorDebug.dll' (Administrado): se cargó 'C:\Program Files
(x86)\Microsoft .NET Micro Framework\v4.2\Assemblies\le\System.dll', símbolos cargados.
'Microsoft.SPOT.Debugger.CorDebug.dll' (Administrado): se cargó 'C:\Program Files
(x86)\Microsoft .NET Micro
Framework\v4.2\Assemblies\le\Microsoft.SPOT.Hardware.SerialPort.dll', símbolos
cargados.
'Microsoft.SPOT.Debugger.CorDebug.dll' (Administrado): se cargó 'C:\Program Files
(x86)\Microsoft .NET Micro Framework\v4.2\Assemblies\le\Microsoft.SPOT.IO.dll', símbolos
cargados.
'Microsoft.SPOT.Debugger.CorDebug.dll' (Administrado): se cargó 'C:\Program Files
(x86)\Microsoft .NET Micro Framework\v4.2\Assemblies\le\System.IO.dll', símbolos
cargados.
'Microsoft.SPOT.Debugger.CorDebug.dll' (Administrado): se cargó 'C:\Program Files
(x86)\Microsoft .NET Micro
Framework\v4.2\Assemblies\le\Microsoft.SPOT.Hardware.PWM.dll', símbolos cargados.
'Microsoft.SPOT.Debugger.CorDebug.dll' (Administrado): se cargó 'C:\Program Files
(x86)\Microsoft .NET Micro
Framework\v4.2\Assemblies\le\Microsoft.SPOT.Hardware.Usb.dll', símbolos cargados.
'Microsoft.SPOT.Debugger.CorDebug.dll' (Administrado): se cargó 'C:\Program Files
(x86)\Secret Labs\Netduino SDK\Assemblies\v4.2\le\SecretLabs.NETMF.Diagnostics.dll',
símbolos cargados.
'Microsoft.SPOT.Debugger.CorDebug.dll' (Administrado): se cargó 'C:\Program Files
(x86)\Secret Labs\Netduino
SDK\Assemblies\v4.2\le\SecretLabs.NETMF.Hardware.Netduino.dll', símbolos cargados.
'Microsoft.SPOT.Debugger.CorDebug.dll' (Administrado): se cargó 'C:\Program Files
(x86)\Microsoft .NET Micro
Framework\v4.2\Assemblies\le\Microsoft.SPOT.Hardware.OneWire.dll', símbolos cargados.
'Microsoft.SPOT.Debugger.CorDebug.dll' (Administrado): se cargó 'C:\Program Files
(x86)\Microsoft .NET Micro Framework\v4.2\Assemblies\le\Microsoft.SPOT.Time.dll',
símbolos cargados.
'Microsoft.SPOT.Debugger.CorDebug.dll' (Administrado): se cargó
'C:\Users\carlos\Downloads\neonmikawebserverV1.2\Framework\NeonMika.Webserver\Ne
onMika.Webserver\bin\Debug\le\..\JSONLib.dll', símbolos cargados.
'Microsoft.SPOT.Debugger.CorDebug.dll' (Administrado): se cargó 'C:\Program Files
(x86)\Secret Labs\Netduino

Netduino 2 en Español

SDK\Assemblies\v4.2\le\SecretLabs.NETMF.Hardware.NetduinoPlus.dll', símbolos cargados.
'Microsoft.SPOT.Debugger.CorDebug.dll' (Administrado): se cargó
'C:\Users\carlos\Downloads\neonmikawebserverV1.2\Framework\NeonMika.Webserver\Ne
onMika.Webserver\bin\Debug\le\..\SecretLabs.NETMF.Hardware.PWM.dll', símbolos cargados.
'Microsoft.SPOT.Debugger.CorDebug.dll' (Administrado): se cargó
'C:\Users\carlos\Downloads\neonmikawebserverV1.2\Framework\NeonMika.Webserver\Ne
onMika.Webserver\bin\Debug\le\..\SecretLabs.NETMF.Hardware.AnalogInput.dll', símbolos cargados.
'Microsoft.SPOT.Debugger.CorDebug.dll' (Administrado): se cargó
'C:\Users\carlos\Downloads\neonmikawebserverV1.2\Framework\NeonMika.Webserver\Ne
onMika.Webserver\bin\Debug\le\..\NeonMika.Util.dll', símbolos cargados.
'Microsoft.SPOT.Debugger.CorDebug.dll' (Administrado): se cargó
'C:\Users\carlos\Downloads\neonmikawebserverV1.2\Framework\NeonMika.Webserver\Ne
onMika.Webserver\bin\Debug\le\..\NeonMika.XML.dll', símbolos cargados.
'Microsoft.SPOT.Debugger.CorDebug.dll' (Administrado): se cargó
'C:\Users\carlos\Downloads\neonmikawebserverV1.2\Framework\NeonMika.Webserver\Ne
onMika.Webserver\bin\Debug\le\NeonMika.Webserver.dll', símbolos cargados.
'Microsoft.SPOT.Debugger.CorDebug.dll' (Administrado): se cargó
'C:\Users\carlos\Downloads\neonmikawebserverV1.2\Executeable\NeonMikaWebserver\Ne
onMikaWebserver\bin\Debug\le\NeonMikaWebserver.exe', símbolos cargados.
'Microsoft.SPOT.Debugger.CorDebug.dll' (Administrado): se cargó 'C:\Program Files
(x86)\Secret Labs\Netduino SDK\Assemblies\v4.2\le\SecretLabs.NETMF.Hardware.dll',
símbolos cargados.
El subproceso '<Sin nombre>' (0x2) terminó con código 0 (0x0).

THANKS FOR USING NeonMika.Webserver
Version: 1.2

Network is set up!
IP: 192.168.1.35 (DHCP: False)
You can also reach your Netduino with the following network name: NETDUINOPLUS

Webserver is now up and running
El subproceso '<Sin nombre>' (0x1) terminó con código 0 (0x0).

ANEXO FINAL

Feedback

Los comentarios de los lectores son siempre bienvenidos. Por favor hágame saber lo que piensa sobre este libro y si es posible, lo que le haya gustado y lo que no, pues es muy interesante conocer su opinión para mejorar futuros proyectos.

Para enviarme comentarios generales, simplemente puede enviarme un correo electrónico a soloelectronicos.es@gmail.com, mencionando el título del libro en el tema de su mensaje.!Gracias!

Fe de erratas

Aunque se han tenido todos los cuidados para asegurar la exactitud del contenido, es muy difícil no cometer errores. Si encuentra un error — tal vez un error en el texto o en el código — le estaría muy agradecido si me informase de ello. Haciéndolo puede ayudar a otros lectores y mejorar las versiones posteriores de este libro.

Si encuentra alguna errata, le estaría muy agradecido que me lo notifique en la cuenta de correo soloelectronicos.es@gmail.com, mencionando por favor el título del libro en el tema de su mensaje. Una vez que haya verificado su fe de erratas, su presentación será aceptada y se tomarán las medidas oportunas para actualizar el contenido.

Preguntas, cuestiones o dudas

Para contactar con el autor puede escribirme a la dirección de correo electrónico soloelectronicos.es@gmail.com, mencionando el título del libro en el tema de su mensaje ,exponiendo en el texto el motivo, la cuestión o la duda que tenga y se hará un esfuerzo para darle una solución.

Sitios de Referencia

Web Oficial de Netduino : http://netduino.com

Web Oficial del producto Microsoft Visual Studio (2010,2012.2013) ; http://www.visualstudio.com

Web de MIT App Inventor: http://appinventor.mit.edu

Web de ARM Cortex :http://www.arm.com/products/processors/cortex-a/

Drivers LCD : http://microliquidcrystal.codeplex.com/

Texas Instruments: http://www.ti.com

Servidor web Neonmika : http://neonmikawebserver.codeplex.com/

Referencias LDR y PIR : http://wikipedia.com

Código fuente aplicación móvil que interactúa con Netduino: http://gallery.appinventor.mit.edu/#page%3DApp%26uid%3D28356001% 26label%3DNetduino%20Control%20%20by%20CRN

Web del autor de este libro donde encontrará mucha información con ideas, montajes y trucos sobre Netduino: http://soloelectronicos.com